Essential Biology

Evolution, Classification
&
Diversity

3rd edition

STERLING
Education

3 2 1

ISBN-13: 979-8-8855717-7-7

Sterling Education materials are available at quantity discounts.

Contact info@sterling–prep.com

Sterling Education
6 Liberty Square #11
Boston, MA 02109

Published by Sterling Education

 Printed in the U.S.A.

STERLING
Education

From the foundations of a living cell to the complex mechanisms of gene expression, *Essential Biology Self-Teaching Guides* are a comprehensive compendium of clearly explained texts to learn and master multifaceted biology topics.

These guides provide a detailed review of essential biological processes of living systems. Develop a better understanding of cell and molecular biology, mechanisms of cell metabolism, plants and photosynthesis, evolution and natural selection, ecology and population biology. Learn the principles of genetics, microbiology, classification and diversity, as well as structure and function of anatomical systems. Reinforce your learning by working through the practice questions and detailed explanations.

Created by highly qualified biology instructors, researchers, and education specialists, these books empower readers by helping them increase their understanding of biology.

We sincerely hope that these guides are valuable for your learning.

230731akp

Essential Biology Self-Teaching Guides

Eukaryotic Cell & Cellular Metabolism

Molecular Biology & Genetics

Nervous & Endocrine Systems

Circulatory, Respiratory & Immune Systems

Digestive & Excretory Systems

Muscle, Skeletal & Integumentary Systems

Reproduction & Development

Microbiology

Plants & Photosynthesis

Evolution, Classification & Diversity

Ecology & Population Biology

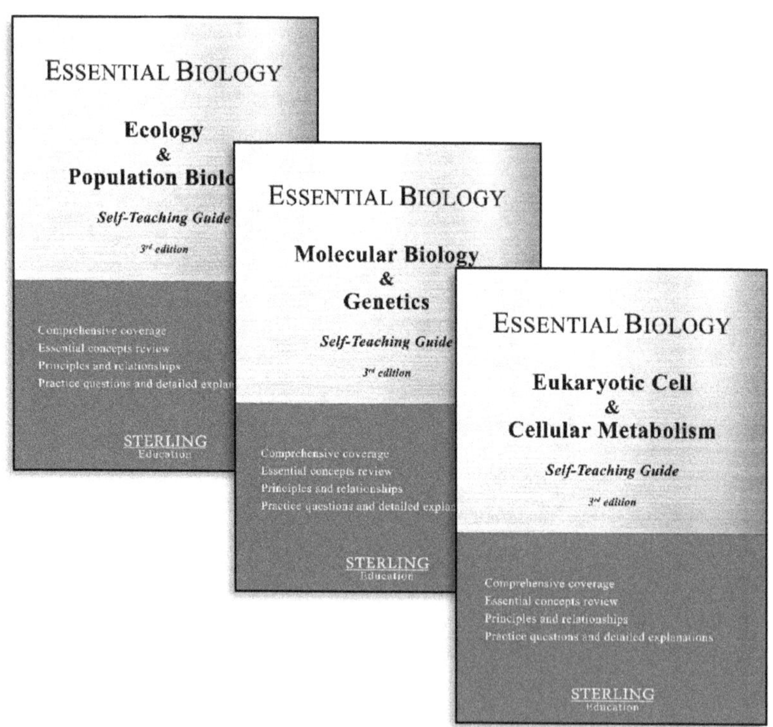

Essential Chemistry Self-Teaching Guides

Electronic Structure & Periodic Table

Chemical Bonding

States of Matter & Phase Equilibria

Stoichiometry

Solution Chemistry

Chemical Kinetics & Equilibrium

Acids & Bases

Chemical Thermodynamics

Electrochemistry

Visit our Amazon store

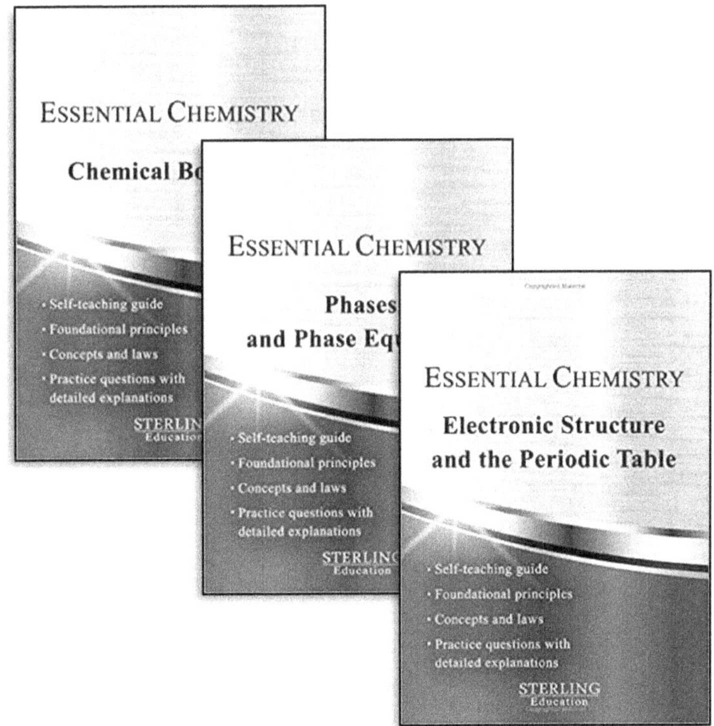

Kinematics and Dynamics

Equilibrium and Momentum

Force, Motion, Gravitation

Work and Energy

Fluids and Solids

Waves and Periodic Motion

Light and Optics

Sound

Electrostatics and Electromagnetism

Electric Circuits

Heat and Thermodynamics

Atomic and Nuclear Structure

Visit our Amazon store

Chemistry

Physics

Cell and Molecular Biology

Organismal Biology

American History

American Law

American Government and Politics

Comparative Government and Politics

World History

European History

Psychology

Environmental Science

Human Geography

Visit our Amazon store

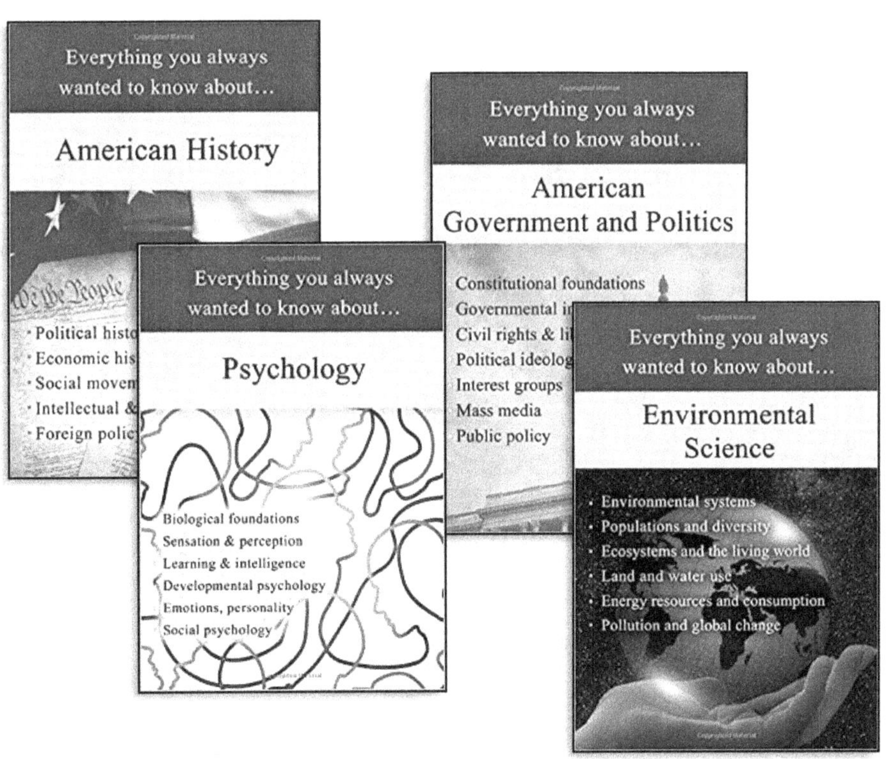

Page intentionally left blank

Table of Contents

Table of Contents (*continued*)

REVIEW: Evolution & Natural Selection (*continued*)

Table of Contents (*continued*)

Table of Contents (*continued*)

Table of Contents (*continued*)

Table of Contents (*continued*)

Table of Contents (*continued*)

REVIEW: Classification & Diversity (*continued*)

Table of Contents (*continued*)

REVIEW

Evolution & Natural Selection

Page intentionally left blank

Foundations of Evolution

Pre-Darwinian worldview

The pre-Darwinian worldview was defined by deep-seated beliefs held to be intractable truths, namely that each species was specially designed and has not changed since Earth's creation.

Variations among same species were explained as circumstantial imperfections in a perfectly adapted creation.

However, Charles Darwin (1809-1882) lived during a profound change in scientific and social realms. His ideas were part of a larger change in scientific thought and perspective.

Georges Louis Leclerc (1707–1788), known by his title Count Buffon, was a French naturalist who wrote the 44-volume *Natural History of All Known Plants and Animals*.

Buffon provided evidence of descent with modification and speculated on how environment, migration, geographical isolation, and competition influence the traits of organisms.

Buffon's work was essential to developing the theory of evolution and natural selection, as he was among the first to assert that traits are *inherited from earlier descendants*.

Paradoxically, Buffon's beliefs often directly contradicted his work, as he believed in a young Earth and the fixity of species.

Common descent theory

Erasmus Darwin (1731–1802), the grandfather of Charles Darwin, was a physician and a naturalist whose writings were on botany and zoology.

Erasmus suggested the possibility of *common descent*, the theory that organisms descended from common ancestors.

Erasmus Darwin based his conclusions on observations of embryonic development and the changes incurred in domestic plants and animals due to selective breeding by people.

Erasmus Darwin offered *no mechanism* by which descent with modification might occur.

Jean-Baptiste Lamarck (1744–1829) was the first to propose *descent with modification* and that organisms adapt to their environments.

Lamarck was an invertebrate zoologist who irked his contemporaries by embracing evolution.

His assertion that organisms change proved correct, but the proposed mechanism was flawed.

Inheritance of acquired characteristics

Inheritance of acquired characteristics was the Lamarckian belief that organisms adapt to their environment during their lifetime and pass on these adaptations to offspring.

He suggested that body parts are enhanced by increased usage while unused parts are weakened.

Lamarck's theory of inheritance of acquired characteristics has been dismissed in favor of Darwin's theory of natural selection and modern additions to the Darwinian theory of evolution.

Lamarck's theory proposed that a giraffe's neck lengthens as it continually stretches to reach leaves, and the slightly lengthened neck would be passed onto the giraffe's offspring.

Over time, these adaptations would accumulate in the long-necked giraffes of modern times.

At the time, however, Lamarck's ideas were fervently held by some scientists, while others, including naturalist Georges Cuvier (1769-1832), adamantly opposed his theory.

The term "evolution" was not used; Lamarck referred to species' changes as *transmutation.*

Darwin's voyage to Galapagos Islands

In 1831, at 22, English naturalist Charles Darwin (1809-1882) accepted a position aboard the *HMS Beagle*; this worldwide voyage provided Darwin with observations shaping his theories.

He spent weeks on the Galapagos Islands, volcanic islands off the South American coast.

Darwin observed island species, which varied much from mainland species and island to island.

For example, Galapagos finches experienced natural selection.

Darwin noted that they resembled the mainland finch but varied in their nesting sites, beak size, and eating habits.

These variations led Darwin to ruminate about the descent of these finches from the mainland species and how isolation on the islands may have contributed to their different traits.

On the Origin of Species

After the *HMS Beagle* returned to England in 1836, Darwin waited over 20 years to publish his findings, knowing the publication would be controversial.

He used the time to publish research and gather further evidence for his grand theory of how life forms arise by descent from a common ancestor and change over time.

In 1859, Darwin felt compelled to publish *On the Origin of Species* when the English naturalist Alfred Russel Wallace (1823-1913) advanced a similar theory.

Foundations of Genetics

Mendelian genetics

Darwin developed his theories in the mid-1800s while Austrian monk Gregor Mendel (1822-1884) studied the genetics of peas.

Mendel cross-pollinated pea plants based on distinctive features to make important discoveries on how traits are inherited between generations.

Darwin may have read Mendel's work but did not include genetics in *On the Origin of Species.*

Mendel's research went unnoticed until his death when early 20th-century scientists realized the profound implications of his experiments.

Newly emerging knowledge about genetics provided an explanation and elaboration on natural selection that Darwin could not describe.

Scientists gradually understood that organisms have a genetic code (i.e., *genotype*), working from the foundations that Mendel laid.

Observable expression of the genetic code (i.e., *phenotype*) depends on gene variants known as *alleles* (i.e., *alternative forms of a gene*).

Hardy–Weinberg equilibrium

On the Origin of the Species popularized *evolution* as the cumulative change of inheritable characteristics within populations, species, or groups.

Evolution occurs if *Hardy–Weinberg Equilibrium,* a tenet of population genetics, is violated.

Hardy–Weinberg Equilibrium states that *phenotypic* (*allelic*) and *genotypic* frequencies *do not change* between generations, provided:

> genes are unaffected by evolutionary forces (*no mutation*),
>
> the population is infinitely large,
>
> without *migration,* and
>
> without *sexual selection* (i.e., mating is random).

In nature, these assumptions are often violated, resulting in evolution.

Darwin described one mechanism of evolution, natural selection.

However, mutation, genetic drift, and gene flow disturb the Hardy-Weinberg Equilibrium.

Genetic drift and gene flow

Genetic drift is random changes in allele frequency in a population and occurs by chance.

Even if they are not the fittest of the population, some organisms pass more genes by chance.

Gene flow is the transfer of genes between two populations.

Gene flow is by migration when a population moves habitat and encounters another population.

Gene flow may be within or between two species for hybridization.

Hybrids are offspring from two genetically dissimilar parents.

Hybrid parents may be from the same or distinct species.

When species hybridize, genes flow between the groups, introducing genetic diversity.

Mutations drive evolution

Mutations are changes in an organism's DNA (genetic code) from damage or replication errors.

Mutations are potent agents of evolution.

Mutations can be fatal, harmless, or beneficial when mutations introduce new traits that allow an organism to succeed; the organism may survive and pass on its mutated code.

Patterns of Evolution

Microevolution and macroevolution patterns

Microevolution describes a change in allele frequency within a population.

The population (known as the *gene pool* or *deme*) is the arena of microevolution.

Microevolution is observed over short periods, especially in rapidly dividing bacteria.

Macroevolution is a pattern of change in groups of related species over long geologic periods. In a sense, macroevolution is the sum of microevolution.

Patterns of macroevolution determine *phylogeny.*

Phyletic gradualism and punctuated equilibrium

Phylogeny is the evolutionary relationship between species and groups of species.

Phyletic gradualism describes the constant, uniform accumulation of small changes, resulting in the gradual transformation of one species into a new species.

Phyletic gradualism is disregarded as a model of macroevolution due to fossil evidence which indicates sudden, drastic speciation.

Punctuated equilibrium, which describes geologically long periods with little to no evolution, is supported by the fossil record, where geologically short periods show rapid evolution.

Gene and genome changes

Random genetic mutations not acted on by natural selection (i.e., environment favoring specific phenotypes) occur constantly.

The amount of elapsed time is determined by measuring the number of neutral mutations.

Genome differences between species can be compared to determine how long ago they may have diverged.

Molecular clocks

Molecular clock dates time comparisons between species.

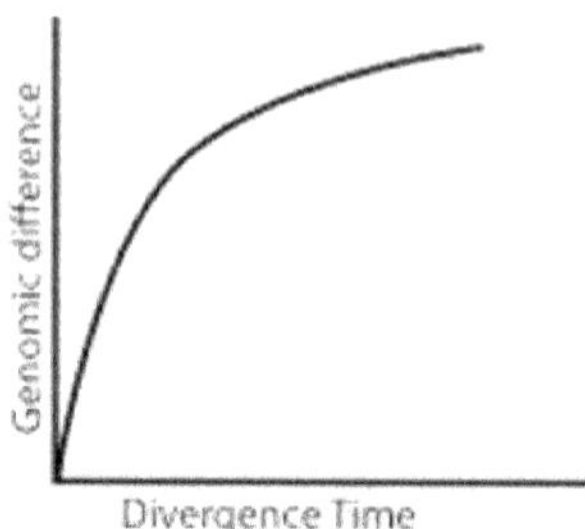

Molecular clock estimates time based on genomic differences

Genetic differences are measured by *phenomes* (i.e., a total of phenotypic traits) or DNA nucleotide sequences.

Molecular clocks indicate relative periods but cannot assign a numerical date.

Calibration of the molecular clock allows for precision *via* comparison against known fossil dates or evidence.

Natural Selection

Survival of the fittest

Natural selection was proposed by Alfred Wallace (1823-1913) and Charles Darwin (1809-1882) as a driving mechanism of evolution.

Wallace and Darwin theorized that environmental pressures select organisms adapted to make the fittest reproduce.

Natural selection has three pre-conditions:

> *First,* population members have random but heritable variations. Individuals differ due to *mutations and chromosomal recombination*; new adaptations arise.

> *Second,* in populations, more individuals are produced in each generation than the environment can support, creating *selective pressure to adapt* to survive in scarce resources.

> *Third,* relying on the prior conditions; some individuals have *adaptive characteristics to survive and reproduce.*

Trait inheritance

Natural selection results in *better-adapted individuals passing adaptations to offspring.*

In contrast, the less-adapted individuals have their alleles eliminated from the gene pool, as they are likely to die before they can reproduce.

Increasing proportions of individuals in succeeding generations have adaptive characteristics.

With time, the population may become a new species altogether.

Since the environment changes, there are no enduring perfectly adapted organisms.

Over time, the fittest organisms may be poorly adapted to a changing environment.

Extinction occurs when the fitness of a population (or species) has declined, whereby in each generation, more individuals die than are born, leading to a dwindling population and the elimination of the group.

Fitness

Fitness is the ability of an organism to pass genetic information on to future generations.

Fitness is a measure of an organism's *reproductive success*.

Evolutionary fitness differs from the colloquial term used to refer to a strong, healthy individual.

Even the healthiest organism is considered "unfit" if it *fails to reproduce.*

Relative fitness compares the fitness of one phenotype to another.

Survival of the fittest is a component of the natural selection theory that describes *how* fitter individuals survive and pass on their traits while fewer fit individuals die out.

Organisms whose traits enable them to reproduce to a greater degree have greater fitness.

For example, black western diamondback rattlesnakes are likely to survive on lava flows, while lighter-colored rattlesnakes survive on desert soil.

Therefore, each species adapts to *maximize its fitness* to survive in its habitat.

Reproduction

Differential reproduction

Survival of the fittest is misleading as it implies that mere organism survival is the driving force of natural selection.

Evolutionary success relies on *reproduction* to pass adaptations to subsequent generations.

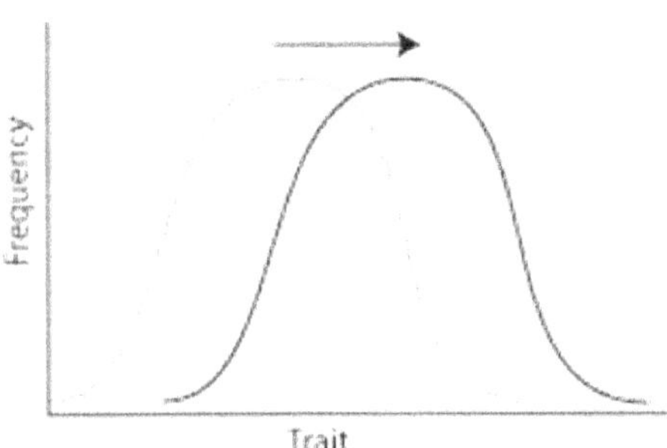

However, organisms may be well-suited for survival but fail to reproduce.

Survival of the fittest may be termed *reproduction of the fittest* or *differential reproduction.*

Differential reproduction links *survival of the fittest* with individuals' *reproductive advantage.*

Differential reproduction occurs naturally, but humans have manipulated natural selection for thousands of years.

Selective breeding

Artificial selection refers to selecting traits in plants and animals that humans prefer, otherwise known as breeding.

Further selection for specific traits resulted in many breeds.

For example, dogs were bred by selecting wolves with friendlier, more domestic traits.

Over time, this artificial selection compounded, producing domesticated dogs.

Domesticated animals and wide crop varieties have been produced by artificial selection.

Breeders of plants and animals try to produce organisms with desirable characteristics, such as high crop yields, resistance to disease, high growth rate, and phenotypical characteristics benefiting people who consume or use these organisms.

Selective breeding often produces a hybrid between two parents with desirable traits.

Hybridization

Hybrid offspring possess desirable parental traits.

For example, one parent may possess dominant alleles for long life, while the other possesses dominant alleles for fast growth.

When *crossed* (or bred), hybrid offspring should be long-lived and fast-growing.

Hybrid vigor is the high fitness characteristic of hybrids.

Hybridization is when two groups with overlapping habitats mate at a geographic boundary.

Hybrid zones are overlapping habitats at a geographic boundary.

Observable Characteristics

Phenotypic variations

Natural selection utilizes random variations (i.e., mutations); therefore, there is no directedness or anticipation of future needs. Organisms are incapable of consciously choosing its genome.

Recessive alleles are expressed in diploid (2N) organisms with two copies (i.e., homozygosity).

Only alleles causing *phenotypic differences* are subject to natural selection.

Heterozygosity preserves recessive alleles

Heterozygosity preserves recessive alleles, as heterozygotes carry rare recessive alleles that may otherwise be selected against.

For example, consider the preservation of two alleles encoding for hemoglobin: a dominant, normal, and recessive, defective allele.

Homozygotes with *both dominant alleles* have typical hemoglobin.

Heterozygotes with expression of one abnormal hemoglobin allele result in the sickle cell trait

Homozygous expression of *both recessive alleles manifests anemia.*

There is a high frequency of sickle cell allele in Africa, with a high malaria incidence.

A link exists between those with sickle cell traits and immunity to the effects of malaria. The sickle cell trait is advantageous in areas where malaria has a higher incidence.

Therefore, a higher frequency of sickle cell alleles in Africa supports natural selection.

There is a tradeoff, as a higher frequency of the allele results in a higher frequency of regressive heterozygosity, causing sickle cell anemia.

Anemia is an excellent defense against malaria, but severe health issues offset this.

Favorable traits

Natural selection is described in several ways.

Directional selection is when a trait at one extreme of a spectrum is favored while traits on the opposite end are selected against. Directional selection shifts the distribution curve of allele frequency toward favored traits.

For example, natural selection leading to drug resistance in bacteria is directional selection.

Another example is trees in the rainforest, competing for sunlight, and selection favors taller trees.

Artificial selection

Artificial selection is mostly directional as humans progressively select traits.

Artificial selection allowed humans to increase the efficiency of livestock animals and crop plants, such as increasing milk yield by continuously breeding cows with high milk production.

Stabilizing selection eliminates extreme phenotypes and favors intermediate phenotypes.

Stabilizing and disruptive selection

Stabilizing selection favors alleles that produce *intermediate* clutch sizes.

For example, Swiss starlings' optimum number of eggs is four or five. If the female lays more or less than this number, fewer survive.

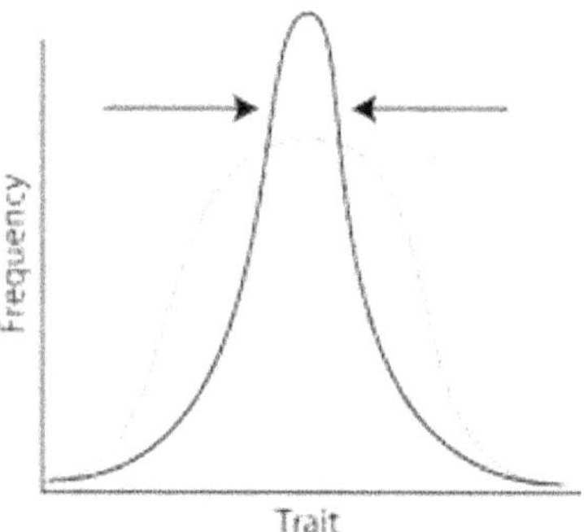

Stabilizing selection favors a narrow range of phenotypes

Disruptive selection occurs when the phenotypes at extremes are favored.

Small beaks are selected for eating berries, while large beaks are selected for cracking seeds.

The intermediate phenotype (medium beaks) is selected against because a medium beak is not helpful for berries or seeds.

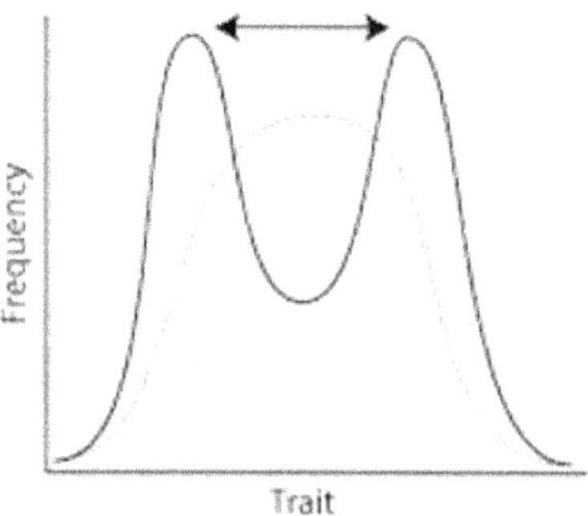

Disruptive selection favors traits at the extremes

Selection Pressures

Group selection

Group selection is a natural selection that acts upon the group and not the individual.

Altruism is when an individual's fitness is sacrificed to benefit the group, usually a family, as the family shares similar genes.

If altruism enables another family member to survive, the genes can be passed on; the individual may exhibit altruism even if sacrificing survival.

Group selection is often provided as an explanation for *altruism*.

Sexual selection

Sexual selection is the differential mating of males (or females) in a population.

In most species, the females select superior males, increasing their offspring's fitness.

Females invest more energy in reproduction and thus attempt to maximize the quality of mates.

Males primarily attempt to maximize the number of their mates.

Male competition for mates leads to fights, with mating awarded to the most potent male.

Mating pressures

Genetically superior males exhibit traits that aid them in this competition and prove their strength (e.g., musculature, large stature, or bigger horns).

Sexual selection by females may lead to traits or behaviors in males that are not practical and do not increase the male's ability to survive.

For example, colorful bird plumages make certain species easily identifiable by predators.

Pressure for males to attract females often leads to *sexual dimorphism*, in which males and females are observably different.

Adaptive radiation

Adaptive radiation is the rapid evolution of many new species from a single ancestral species.

Adaptive radiation occurs when the ancestral species is introduced into an area where diverse geographic or ecological conditions are available for colonization.

For example, observations by Darwin of finches illustrate the adaptive radiation of several species from one founder species of mainland finch.

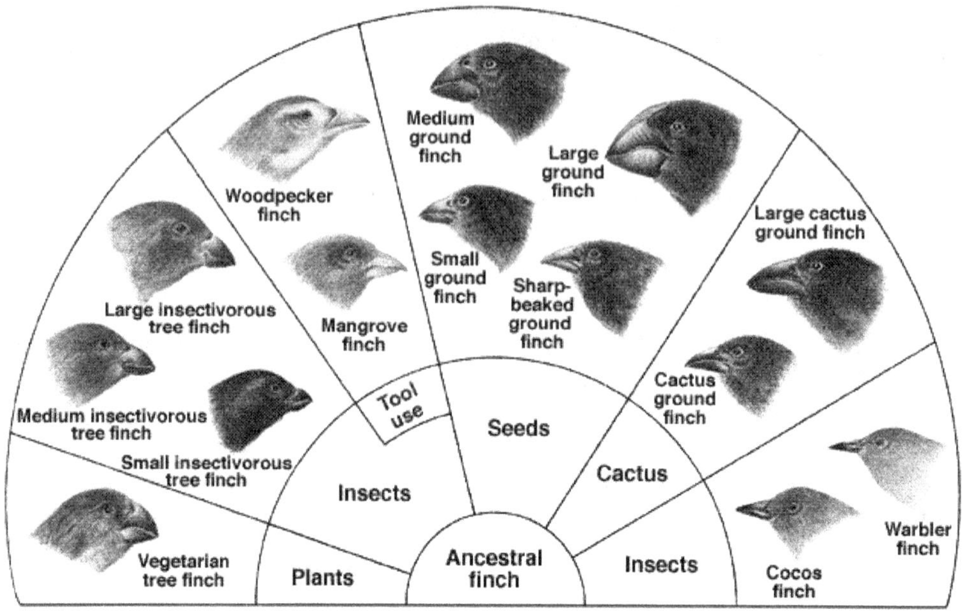

Darwin's theory of finch evolution on the Galapagos Islands
illustrates adaptive radiation from differences in their environments

Evolutionary success increases gene pool

Increase in allele frequency represents evolutionary success for that allele.

Increase in an individual's allele frequency is an evolutionary success for that individual.

For example, the peppered moth in European cities illustrates changes in allele frequency.

Pre-industrialization, these moths were light-colored. However, as pollution increased in the 1800s, soot collected on the sides of buildings, making the light-colored peppered moth more visible to predators.

Mutations resulted in a new phenotype, darker than the white peppered moth.

Darker peppered moths were favored over lighter moths by natural selection, specifically directional selection.

Allele frequency changes

Over time, the frequency of the dark allele increased in populations, reaching 95% in most industrial European cities.

However, soot has been reduced, and most buildings are lighter than in the late 1900s.

Therefore, the allelic frequency of dark-peppered moths has decreased since, demonstrating *directional selection back* towards a light phenotype.

Speciation

Phenotypic and genotypic changes

The definition of a species is controversial and inexact.

At a particular stage of evolution, one population may become so genetically different from the other that it becomes a new species, the process of *speciation*.

After enough changes in the gene pool, phenotypic and genotypic frequencies have amassed, and *speciation is inevitable*.

However, distinct species cannot reproduce and produce viable and fertile offspring.

Biogeography

Biogeography is the study of the geographic distribution of life forms on Earth.

Comparing animals of South America and the Galápagos Islands led Darwin to conclude that environmental adaptation can cause diversification, including the origination of new species.

Darwin's conclusions were influenced by Charles Lyell (1797-1875), a geologist who presented the idea that geological variations were formed by slow, continuous processes such as erosion.

This hypothesis contrasted with the prevailing belief that Earth's contours had been shaped by divine intervention and had only changed due to sudden, violent catastrophes.

Lyell's theory of *uniformitarianism* supported Darwin's observations of geology and fossils.

Plate tectonics

Lyell and Darwin's conclusions were the precursors to *tectonic plate theory*, which describes the movement of Earth's crust.

In 1920, German meteorologist Alfred Wegener (1880-1930) presented data supporting the highly controversial *continental drift* model.

By the 1960s, continental drift was confirmed by strong geologic and evolutionary evidence.

For example, many of the same fossils have been found on different continents, indicating a time when the continents were united.

Continental drift is why the coastlines of several continents mirror images of each other, as in the outlines of the west coast of Africa and the east coast of South America.

Geological features exist throughout regions where the continents separated or collided.

Pangea

Pangea was a single ancestral supercontinent separated over millions of years into land masses.

As the continents drifted apart, organisms were separated, resulting in great speciation.

One example is the marsupials of Australia, who were isolated from other mammals, and therefore are different from mammals on other continents.

Allopatric and sympatric speciation

Allopatric speciation results in populations separated by geographic barriers.

While geographically isolated, variations accumulate via natural selection, mutation, gene flow, and genetic drift until the two populations are reproductively isolated.

Sympatric speciation is when a single population member, without geographic isolation, develops genetic variations preventing reproduction with the original population.

For example, sympatric speciation is observed in polyploidy in plants.

Polyploidy organisms

Polyploidy possesses more than two sets (2N) of chromosomes (e.g., 3N, 4N).

Failure to reduce the chromosome number produces polyploid plants that reproduce successfully only with other polyploid plants.

Backcrosses with diploid plants are sterile. Polyploidy leads to new species of plants.

Aside from polyploidy, evidence for sympatric speciation is sparse; the issue is under intense research and debate in evolution.

Researchers contend that sympatric speciation is a subtype of *allopatric speciation*, and interbreeding makes sympatric speciation impossible without a *geographic barrier*.

Classifications of Organisms

Taxonomy

Taxonomy classifies organisms, an undertaking attempted for thousands of years.

Carl Linnaeus (1707–1778), a Swedish naturalist, revolutionized taxonomy by introducing a streamlined, standardized taxonomic system.

Linnaeus' taxonomic system replaced hundreds of previously used disorganized systems.

Linnaeus' system used binomial nomenclature with a two-part name for each species (e.g., *Homo sapiens*).

Divine hierarchy of life

Like other taxonomists, Linnaeus believed each species had an "ideal" structure and held a fixed place in the *scala naturae*, the *divine hierarchy of life*.

Linnaeus thought that classification should describe the fixed features of species and reveal God's plan and believed that the ideal form of each organism could be deduced and arranged according to the *scala naturae*.

Linnaeus was the first prominent scientist proposing humans are closely related to primates.

This bold declaration clashed with humans being fundamentally distinct from animals.

Linnaeus' later work with hybridization suggested that *species might change with time*.

Taxon and species

Classification establishes categories of species based on their relationship to other species.

Taxon is a group of organisms in a classification category; *Homo* or *Felis* are taxa at a genus level, describing the genera of hominids and cats, respectively.

Species are a taxonomic category below the rank of *genus*.

When a species has a wide geographic range, variant types may interbreed where they overlap; these populations are *subspecies*.

For example, *Canis lupus* contains around 40 subspecies, including *Canis lupus lupus* (Eurasian wolf) and *Canis lupus familiaris* (domestic dog).

Adding the subspecies makes for a trinomial or three-part name.

Classification changes

Taxonomic classifications separated organisms into two categories, Animalia and Plantae.

This is incorrect; many organisms were discovered fitting both (and neither) categories.

Taxa are kingdoms, phyla, classes, orders, families, genera, and species from the broadest.

Super-, sub-, or infra add classifications to taxonomic levels.

Currently, at least seven taxonomic levels exist.

Domain is a taxon above kingdom and is classified into Eukarya, Bacteria, and Archaea.

Domains and subsequent taxa are under revision, controversy, and debate.

Genetic Factors of Speciation

Defining species

Contentions in taxonomy come from attempts to define species.

Controversy regarding species criteria has sparked books and movements and has been ongoing since the earliest days of taxonomy.

Many arguments involve philosophy, making consensus incredibly difficult.

Biological definition of species is a group of organisms that may interbreed and produce fertile, viable offspring without outside intervention.

Gene flow via sexual reproduction cannot occur between reproductively isolated species.

Reproductive isolation

Reproductive isolating mechanism is any structure or behavior preventing reproduction.

Prezygotic isolating mechanisms prevent successful mating or fertilization.

Prezygotic barriers include habitat and temporal, behavioral, and mechanical isolation.

Habitat isolation is when species occupy different habitats and are less likely to meet and attempt to reproduce.

Temporal isolation is a *prezygotic* barrier where species occupy the same habitat but mate in different cycles (i.e., times of the year).

Behavioral isolation

Species may have *behaviors* precluding mating, such as *behavioral isolation.*

For example, flowers exhibit behavioral isolation when one is bee-pollinated and the other bird-pollinated so that the two do not exchange gametes.

Another example is species with different mating rituals so that females of one species do not consider males of the other species as mating partners.

Even if males and females mate, *mechanical isolation* may prevent successful mating due to incompatible reproductive structures or anatomy.

Genetic isolation

Gamete isolation is prezygotic reproductive isolation from the incompatibility of gametes from two species so they cannot fuse to form a zygote (or *fertilized egg*).

For example, eggs may have receptors for species-specific sperm.

Postzygotic isolating mechanisms prevent the development of a *hybrid* (i.e., offspring from two species or varieties) after fertilization.

Postzygotic isolating barriers include hybrid sterility or inviability.

Hybrid breakdown

Hybrid breakdown is when hybrids are fertile in the first generation, but the second generation is infertile.

Populations may reunite before reproductive isolation is complete; they may reproduce fertile hybrids with increased genetic diversity, which benefits the population.

Defining species from reproductive isolation has limitations.

Some groups are considered separate species from other differences, even if they can hybridize where their ranges overlap.

Furthermore, many species do not reproduce sexually and thus cannot be characterized by sexual reproduction abilities.

Overall, it is difficult to test and observe reproductive isolation.

Sexual dimorphism

Polymorphism is *genetic variation* (*alleles*) within a population for *natural selection*.

Early taxonomists such as Linnaeus classified organisms based on morphology, but this method has several limitations.

This becomes apparent for variations among the same species, like *sexual dimorphism*.

Sexual dimorphism exemplifies *morphological variations* within a species or *polymorphism*.

Morphisms

Phenotypic polymorphism

Phenotypic polymorphism exists in two or more distinct *phenotypic variations* within a species.

Sexual dimorphism falls under phenotypic polymorphism, as does blood ABO type.

Determining whether two groups are polymorphs of the same species or different subspecies may be challenging.

Polymorphism may result from sexual selection (e.g., *sexual dimorphism*) or ecological niches.

The two different forms of the peppered moth discussed earlier exemplify *polymorphism*.

Genetic polymorphism

Genetic polymorphism applies the concept of polymorphism to the genotype.

In genomics, genetic polymorphism refers to the *unique differences* between the genomes of two individuals. In this regard, a species has almost infinite *genetic polymorphs*.

Balanced polymorphism is natural selection favoring the balance of polymorphs in populations; the frequencies for each allele are equal.

Heterozygote advantage

Heterozygote advantage is when the heterozygous expression of two alleles is favorable; it is one mechanism of balanced polymorphism.

For example, conserving recessive sickle cell alleles in populations afflicted by malaria illustrates this mechanism, as heterozygotes have an advantage over homozygous individuals.

Balanced polymorphism

Balanced polymorphism arises by *frequency-dependent selection* when one phenotype's fitness depends on another; an increase in the frequency of either phenotype is balanced by the other.

In contrast to genetic polymorphism, environmental effects may control polymorphism by a *polyphyletic system* (i.e., *phenetic polymorphism*).

Phenomes

An individual's *phenome* (i.e., phenotypes expressed), rather than the genome, is altered to produce a morph (or *phenotypic difference*).

For example, many organisms have *polyphyletic* (i.e., derived from more than one common evolutionary ancestor) sex determination.

Sexual development is regulated by factors such as temperature, predator density, or the current male-to-female ratio in populations.

When polyphenic forms balance, they are challenging to distinguish from genetic polymorphs.

Environmental Factors of Speciation

Adaptation and specialization

Adaptation is when an organism evolves to become suited to its environment.

Because of natural selection, adaptive traits accumulate in each succeeding generation.

Specialization occurs as adaptations allow species to exploit a niche.

Organisms adapt in response to the evolution of other species within their habitat.

Most environments change, and so organisms *adapt* or become *extinct*.

Organisms may adapt to habitat environmental changes (e.g., soil acidity or precipitation); this new habitat may be from migration or an invasion from outside species.

Co-adaptation

Co-adaptation is when species evolve in response to another.

For example, pollinating insects and flowers continually co-adapt to maximize their mutualistic relationship.

Co-adaptation may be more threatening when a host species evolves into a hostile gut environment; the parasitic species must adjust to survive in the host species' gut.

Exaptation is when, over time, many traits are readapted for a new purpose.

Maladaptation is when traits are selected against because they become a distinct disadvantage.

Ecological niches

Niche encompasses both physical and environmental conditions.

Ecological niche is the environment in which an organism lives and its role.

Ecological niche is the organism's *habitat* (i.e., *area and resources used*), behavior, and relationship with the habitat.

Organisms attempt to fill their niche by *maximizing resources*.

The specialization of the organism accomplishes this as it adapts to its ideal niche.

Competition

Competition is an intense *selection pressure* driving organisms to evolve and adapt to a different niche or better compete for resources in their current niche.

Resources are best utilized when organisms occupy distinct niches, as they do not compete.

When niches overlap, resources become scarce, and organisms threaten each other's survival. This results in competition, which can be *interspecific* and *intraspecific*.

Intraspecific competition is typical since members of the same species occupy similar niches.

Population growth is balanced by competition.

Competition is minimal at a low population density, and population growth can occur rapidly.

However, as the population grows, the available resources dwindle.

Carrying capacity

Carrying capacity describes the maximum population that the environment can support.

Populations at carrying capacity, *intraspecific competition increases,* and *populations slow*.

Competition within a species can force members to occupy different niches, driving speciation.

Individuals may compete for mates, food, water, space, or other advantages.

They compete indirectly by *depleting resources* that other organisms also utilize.

Competitive exclusion principle

Competitive exclusion is when species use the same resource in the same place, diverging into different niches by *niche differentiation*.

If this divergence does not occur quickly, one species outcompetes the other by pushing the species from the shared overlap of their niches.

Exponential population growth is rare due to the pressures of competition.

Humans exhibit exponential population growth by finding and exploiting new resources.

Population Growth Strategies

r-selection growth

Evolutionary ecology places species on a spectrum related to their population growth strategy.

r-selected species mature rapidly, reproduce early, and produce many offspring.

A classic example of an r-selected organism is the mouse.

Mice populations grow exponentially due to *early reproductive capacity* and *large broods*.

Mice preserve energy otherwise spent caring for young by minimizing parental investment; this results in a high offspring mortality rate.

Mice populations quickly exceed the environment's carrying capacity, exhausting resources.

Exhausting resources result in sudden and intense competition (along with disease, overcrowding, and other threats), and the *population dwindles*.

Fluctuations in population growth and subsequent decline typify r-selected organisms.

K-selection growth

K-selected organisms, such as primates, tend to mature slowly.

K-selected organisms usually maintain a *steady population* near their carrying capacity.

Once at reproductive age, they produce few offspring but reproduce throughout life.

They live longer and have a higher offspring survival rate but require great energy investment.

Organisms frequently exhibit r-selected traits in some areas and K-selected traits in others; often between r- and K-selection and shifting strategies depending on environment and density.

Inbreeding

Inbreeding is mating between closely related individuals.

Inbreeding increases the frequency of homozygotes, decreases the frequency of heterozygotes, and decreases genetic diversity.

Some organisms avoid inbreeding because it *increases disease rates* and other undesirable traits. However, inbreeding often occurs in nature, being unavoidable in small populations.

Inbreeding is a typical result of *artificial selection* by humans.

Inbreeding depression

Inbreeding for selecting traits results in:

loss of specific genes

decreased genetic diversity

accumulation of genetic defects

Inbreeding depression is the loss of fitness resulting from inbreeding effects.

It is generally advantageous for organisms to promote genetic diversity.

Present-day breeders strive to avoid inbreeding.

Genetic diversity

Genetic diversity serves as a way for populations to adapt to changing environments.

Variation in the gene pool may help a species be prepared for a wide range of scenarios, such as food shortage or an epidemic of disease.

For example, an extremely contagious disease may threaten 99% of a species, but 1% possess an allele that provides disease resistance.

With this allele, the population can survive when otherwise extinct.

Outbreeding increases population fitness

Outbreeding is genetically dissimilar individuals mating to increase genetic diversity.

Outbreeding (*vs.* inbreeding) increases the *fitness of a population*.

However, crossing individuals from different populations rarely results in less fit offspring than within the same population. This is *outbreeding depression*.

Outbreeding depression may result from disruptive selection when the *heterozygous genotype* is selected *against* in favor of the homozygous genotype.

Outbred progenies display an intermediate phenotype that is not useful in the populations in which their parents originated.

Bottlenecks

Bottleneck is a severe *reduction in population size.*

Bottlenecks can be deleterious to the population because the population is less able to adapt to environmental changes with a smaller, less diverse gene pool.

For example, bottlenecks can be caused by a natural disaster eliminating most of the population.

Bottlenecks increase the *effects of genetic drift* (i.e., random fluctuations in allele frequencies) because traits may be represented disproportionately when a population drastically decreases, especially if the decrease is random.

For example, a natural disaster may kill mostly brown rabbits randomly in a population of white and brown rabbits in equal numbers.

Bottlenecked populations display a disproportionate representation of the white rabbits' alleles. The population may grow in subsequent generations, but primarily white rabbits.

Genetic drift and founder effect

Genetic drift is the random fluctuations in allele frequencies in a population; amplified when a small group migrates out of a large population.

Founder effect, as with a bottleneck, is when a small population has a limited gene pool that may not accurately represent the original population.

Founders can profoundly affect the population's gene pool after generations.

Populations diverge from genetic drift with:

> *mutations* (i.e., *nucleotide sequence changes*)

> *gene flow* (i.e., *allele transfer between populations*)

> *natural selection* (i.e., *environment-favoring phenotypes*)

Notes for active learning

Evolutionary Progressions

Divergent evolution

Divergent evolution is when species from the same *lineage* (*common ancestor*) evolve to be increasingly different.

For example, bats and horses share a mammalian lineage, but the ancestral mammalian forelimbs became wings in bats and hooves in horses.

Homologous structures arose from the same ancestor (e.g., bat's wings and horse's hooves).

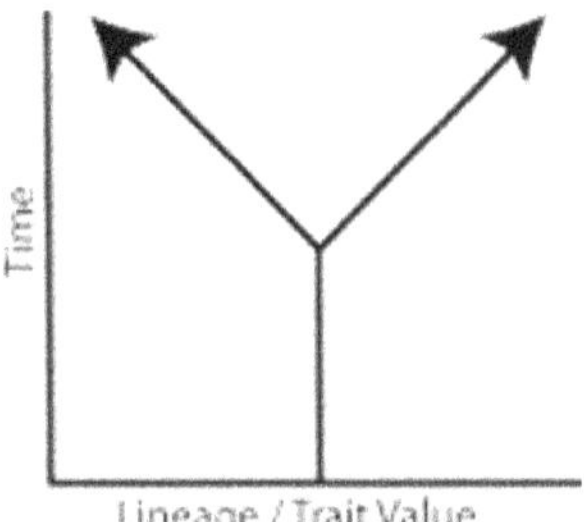

Divergent evolution gives rise to homologous structures

Parallel evolution

Parallel evolution is when two related species of the same lineage evolve similarly.

For example, feeding structures of different crustacean species originated from ancestral leg mutations, evolving them into mouthparts.

Parallel evolution is same lineage, traits, and evolution from similar mechanisms or mutations.

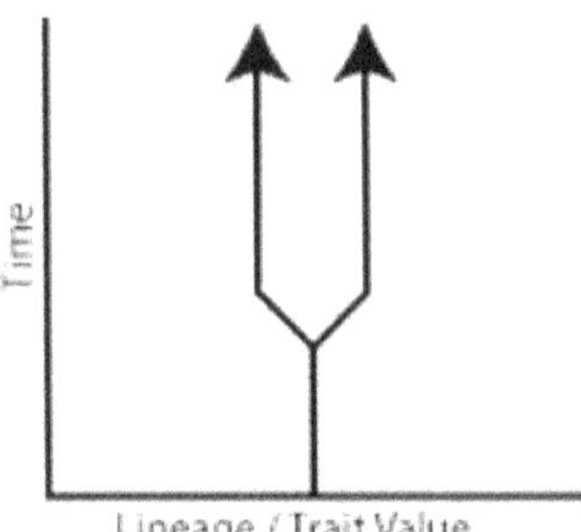

Parallel evolution for two related species undergoing a similar evolution

Convergent evolution

Convergent evolution is unrelated species of different lineage, by different mechanisms, evolving similarly.

For example, bats and butterflies have wings but came from different lineages and evolved through different mechanisms or mutations.

Analogous structures are similar but arose by convergent evolution.

Analogous structures can be deceptive, as they often suggest common lineage when the structures are similar merely *by chance*.

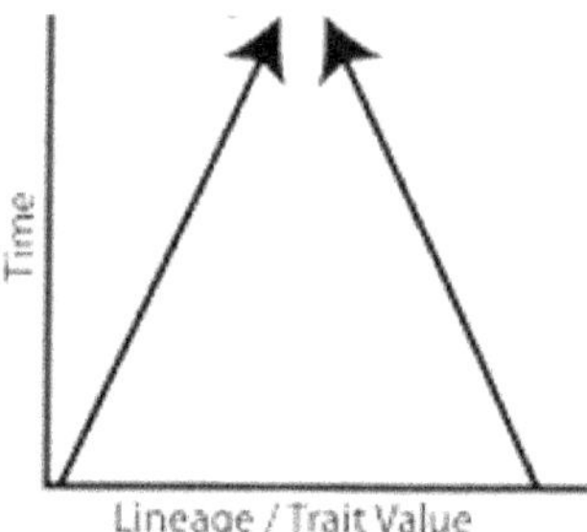

Convergent evolution from unrelated species that become similar

Coevolution

Coevolution is when species evolve in response to one another, accumulating co-adaptations.

For example, a predator may develop a trait that aids in hunting prey.

In response to this newly developed trait in the predator, the prey may evolve a trait allowing it to evade the predator more successfully.

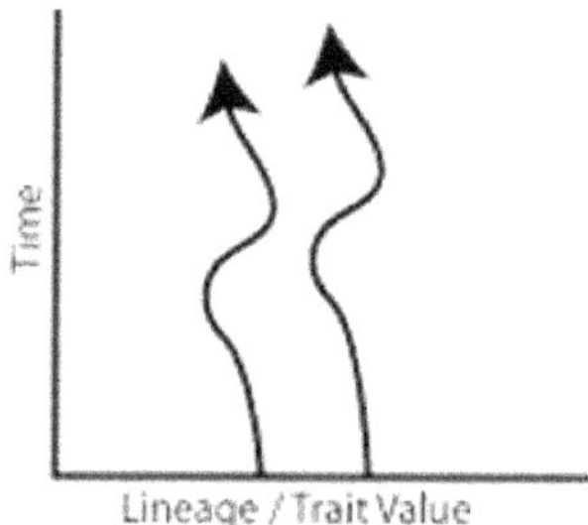

Coevolution with an evolutionary relationship due to another species

Origin of Earth

Big bang theory

About 13.8 billion years ago (b.y.a.), all material in the universe was created in a massive expansion of matter during the *Big Bang*.

Matter and space were suddenly expelled from a *singularity of infinite density*.

The sudden cooling of the superheated ejecta facilitated the combining of atomic components into atoms and molecules.

Gas clouds cooled and formed the principal components of galaxies, including stars and planets.

Earth formed about 4.6 b.y.a. when dust and gasses left from the formation of the sun coalesced.

Heat and gravity stratified the Earth into several layers:

molten core of iron and nickel,

semi-liquid mantle of silicate minerals, and

solid crust created by the upwelling of lava from the mantle.

Primordial atmosphere

Early Earth was so hot that water was vapor in dense, thick clouds.

Earth cooled for 1 billion years, water vapor condensed into rain, forming primordial oceans.

Earth's *size* provides a *gravitational field* strong enough to hold an *atmosphere*.

Gases from volcanoes and celestial bodies bombarding Earth formed the *primitive atmosphere*.

Earth's primitive atmosphere differed from its current atmosphere, which consists of

water vapor (H_2O)

nitrogen (N_2)

carbon dioxide (CO_2)

lesser amounts of hydrogen (H_2) and carbon monoxide (CO)

Abiotic (non-living) factors are physical and chemical components (light, temperature, pH).

Oxidizing *vs.* reducing atmosphere

The primitive atmosphere contained *little free oxygen* (O_2) and was a *reducing atmosphere*, as opposed to the oxidizing atmosphere of today.

Oxidizing atmosphere contains free O_2 and inhibits complex organic molecule formation.

Reducing atmospheres lack free O_2 and form complex organic molecules (i.e., carbon-based).

Primitive atmosphere was crucial for simple elements forming essential organic molecules.

Another significant contribution may have come from the comets and meteorites that pelted the Earth, carrying organic chemicals.

Evidence suggests that the primitive atmosphere was less reducing than previously thought, with less ammonia.

Abiotic synthesis may have evolved in oceans at *hydrothermal vents*, which expelled substantial amounts of *ammonia* (NH_4).

Chemical evolution may have occurred at these deep thermal vents and in the atmosphere.

Origin of Life

Oparin's hypothesis

Chemical evolution is the increased complexity of chemicals leading to the first cells.

Chemical evolution gave way to biological evolution when self-replicating molecules arose and underwent natural selection.

Like organisms, biotic molecules undergo variations and natural selection, enabling evolution.

In the early 1900s, Soviet biochemist Alexander Oparin (1894-1980) proposed that organic molecules formed from elements of a reducing atmosphere:

methane (CH_4), ammonia (NH_3), hydrogen (H_2), and water (H_2O).

In Oparin's vision, lack of oxidation and decay allowed organic molecules to form a thick, warm organic soup from which life arose.

Oparin's theory motivated chemists, such as Stanley Miller, to perform experiments that imitated the conditions of the early Earth.

Urey–Miller experiment

In 1953, Stanley Miller (1930-2007), supervised by Nobel laureate Harold Urey (1893-1981), published results from experiments on the chemical origin of life, the *Miller Spark-Discharge experiment,* or the *Urey–Miller experiment.*

Miller's experiment used a reducing atmosphere bombarded by kinetic energy from lightning.

A sealed flask of water vapor simulated the early atmosphere; methane, ammonia, and hydrogen were continually subjected to electrical sparks.

A deep red solution formed hydrogen cyanide (HCN), formaldehyde (CH_2O), and related compounds within a week.

Simple organic molecules assembled into different amino acids, the monomers of proteins.

Amino acid formation

Miller concluded that preliminary organic molecules originated in a reducing atmosphere.

Amino acids formed readily in Miller's spark-discharge experiment to have formed in the conditions of early Earth.

Recent evidence suggests that amino acids could have been seeded from comets and meteorites.

Interstellar dust constantly falling on Earth contains hydrogen cyanide and aldehydes, key reactants forming amino acids.

Evidence since Miller's experiment has indicated that Earth had a more oxidizing atmosphere than previously thought.

Atmospheric composition

Current scientific thought proposes an atmosphere of 60% hydrogen, 20% water vapor, 10% carbon dioxide, and 5% hydrogen sulfide.

Experiments based on this assumption yield more diverse molecules than those from the Urey-Miller experiment, including amino acids, nitrogenous bases, sugars, and hydroxy acids – all crucial for the existence of life.

One puzzle that persists is that chemical evolution experiments have yielded a mixture of *chiral* left-handed and right-handed *enantiomers* (*non-superimposable mirror images*).

Although both chiral forms of every amino acid are in nature, only left-handed forms are in living organisms.

Scientists have not yet answered how left-handed forms were utilized in living organisms.

Polymerization Hypotheses

Polymers evolved

Polymerization occurs by condensation, which removes a water molecule (i.e., *dehydration*) from two monomers to form a bond.

Hydrolysis (*lysis by water*) depolymerizes macromolecules with H_2O to *break bonds*.

Earth's primordial oceans collected organic monomers formed in the atmosphere, and amino acids and nucleotides polymerized and formed proteins and nucleic acids.

Hydrolysis and polymerization reactions are two challenges in forming polymers.

Firstly, hydrolysis dominates the chemical equilibrium over condensation because hydrolysis increases entropy (*randomness*) while condensation decreases it.

Therefore, breaking polymers by hydrolysis is thermodynamically (*energetically*) favorable.

Polymerization

Polymerization requires energy input by coupling the reaction with an energy-producing reaction (e.g., hydrolysis of ATP).

However, no mechanism has been widely accepted for this in the pre-biotic era; it is challenging to model polymerization reactions based on the presumed conditions of early Earth.

Scientists recently showed that macromolecules could have polymerized in early Earth's muddy tide pools and beaches.

They mixed nucleotides and minerals from mud and collected the resulting polymers, isolating nucleic acids up to 40 nucleotides long.

The experiment was repeated with amino acids, and once again, they found that the monomers had managed to polymerize into proteins up to 55 amino acids long.

Protein World Hypothesis

Protein World Hypothesis, or *Protein-first Hypothesis,* proposed in the 1950s by Sidney W. Fox (1912-1998), demonstrated amino acids form peptides abiotically if exposed to dry heat.

This hypothesis was that amino acids collected in shallow puddles along rocky shores and the sun's heat caused them to form *proteinoids.*

These small polypeptides have some catalytic properties.

Proteinoids form tiny *microspheres* in water due to hydrophobic amino acids.

Proteinoid microspheres are like lipid cell membranes, so the Protein World Hypothesis contends that *polypeptides were the precursors* to living entities.

This hypothesis has been refuted since it was determined that *proteinoids are not proteins.*

Furthermore, the first living macromolecules would need to be able to self-replicate, which neither proteinoids nor proteins can.

RNA World Hypothesis

RNA World Hypothesis states that RNA was the precursor of the first cell.

In 1982, Thomas Cech (b. 1947) and Sidney Altman (1939-2022) demonstrated that unique RNA called *ribozymes* function as enzymes; they were awarded the Nobel prize in 1989.

Some viruses have RNA genes with enzyme reverse transcriptase, using RNA to form DNA.

This evidence shows that RNA possibly assumed the functions of DNA and protein enzymes on early Earth, leading to both.

Supporters of this hypothesis label this an "RNA world."

RNA is an excellent model for chemical evolution.

Inexact copies (mutations) are made when replicating, which produces diversity in the population of self-replicating RNA.

Mutants are efficient self-replicators with more copies of themselves, dominating the population.

Less efficient self-replicators may be destroyed by UV radiation or chemical reactions.

Thus, chemical evolution changes the composition of the population.

This is natural selection, directed toward abiotic systems rather than living organisms.

Most researchers subscribe to an RNA world, but this is not a definitive theory.

Clay hypothesis

Clay hypothesis was proposed by Graham Cairns-Smith (1931-2016) in the 1960s, which proposes that amino acids polymerized in clay, with radioactivity providing energy.

Clay attracted organic molecules and contained iron and zinc atoms, which were inorganic *catalysts* for forming polypeptides (*proteins*) and biomolecules.

Hypothetically, clay collects energy from radioactive decay and discharges it during temperature or humidity changes.

Nucleotides and amino acids may have polymerized in this environment and associated, which evolved simultaneously.

An advantage of the clay hypothesis is that it addresses the *chicken-and-egg paradox*, with RNA and proteins arising simultaneously.

DNA is not a viable candidate for the first living entity because it has low reactivity (i.e., not *labile*) and is not an effective catalyst.

Unlike RNA, DNA has never been shown to have catalytic properties.

Iron-Sulfur World Hypothesis

Several hypotheses operate under the assumption that life began at *hydrothermal vents*.

Iron-Sulfur World Hypothesis (a chemosynthetic model) suggests that life arose in hydrothermal vents, aided by the high pH, temperature, and richness of iron and sulfur.

Iron-Sulfur World Hypothesis proposes that lipids were synthesized and anchored to the vent walls before forming a lipid membrane sphere and associating with proteins to form a *protocell*.

In 1998, Gunter Wachtershauser (b. 1938) and Claudia Huber described how they created peptides using iron-nickel sulfides under vent-like conditions.

Minerals have a charged surface, attracting amino acids and providing electrons to polymerize.

Zinc World Hypothesis is a hydrothermal vent model focused on zinc rather than iron and sulfur.

Notes for active learning

Protocells

Evolution of protocells

Before the first true cell arose, there was a *protocell* with a lipid-protein membrane and the ability to perform metabolic reactions.

Although the concept of the proteinoid has been paused, Fox's experiments proved that lipids and proteins could form complex membranes under the right conditions.

Coacervate droplets are spheres of lipids that spontaneously form under appropriate temperatures, ionic environments, and pH.

Coacervate droplets selectively absorb and incorporate substances from the surrounding solution, like the *semi-permeable membranes* of cells.

Oparin believed that the protocell could have developed from coacervate droplets.

The definitive model for the protocell is not recognized.

Oparin's hypothesis of protocells from coacervate droplets is relevant, but some concepts have been superseded.

Micelles

Some scientists postulate that fatty acids organize into *micelles*, another form of a lipid sphere, in a mineral-rich clay environment.

Clay allows for the development of nucleotides that spontaneously assemble into RNA.

Abiotic amino acid synthesis and peptide polymerization could have occurred at *deep-sea hydrothermal vents*.

The first protocell may have used pre-formed ATP to drive reactions, but natural selection would favor protocells that could synthesize ATP as supplies dwindled.

Chemosynthetic and thermosynthetic models

Chemosynthetic models propose that energy for peptide synthesis comes from electrochemical gradients within the microcompartments of the vents.

Chemiosmosis is integral to cellular respiration and photosynthesis, describing the movement of hydrogen atoms across a membrane to drive the enzyme ATP synthase.

Thermosynthetic models suggest thermal cycling as the primary energy source: thermosynthetic models posit chemiosmosis, not glycolysis, as a method of ATP production.

Thus, thermosynthesis assumes ATP synthase was the first protein enzyme to evolve.

Fermentation and glycolysis

Both hypotheses assume that early energy production was accomplished by fermentation, which generates ATP without oxygen; glycolysis was a crucial adaptation for protocells.

Glycolysis pathway is prevalent in living organisms, indicating that it evolved early in life.

Autotrophs make organic molecules from inorganic nutrients.

Heterotrophs cannot synthesize organic compounds from inorganic substances and must intake organic compounds.

If a protocell were a *heterotrophic fermenter* living on organic molecules in the *prebiotic soup*, this would indicate that *heterotrophs preceded autotrophs*.

However, *chemosynthetic autotrophs* would have preceded heterotrophs if the protocell evolved at *deep-sea hydrothermal vents*.

Chemical Evolution

Self-replication

Chemical evolution theory rests on the idea that the precursor to life was a self-replicating molecule, which many agree would have to be RNA.

Protein-first Hypothesis supporters argue that proteins could have self-replicating abilities that have not yet been discovered.

Genetic information flows from DNA to RNA to protein in most biological systems (i.e., the *central dogma of molecular biology*).

DNA $\rightarrow$ RNA $\rightarrow$ protein path may have developed in stages.

Once the protocell could reproduce, it became a true cell and began biological evolution.

After DNA formed, the genetic code evolved to information storage.

Current genetic code has fewer errors, minimizing mutations; it underwent *natural selection*.

Most biologists suspect life evolved in basic steps.

Abiotic synthesis of organic molecules such as amino acids occurs in the atmosphere or at hydrothermal vents.

Polymers and protocells

Monomers joined to form polymers, perhaps in *tidal pools, clay,* or *hydrothermal vents.*

First polymers were likely *RNA,* followed closely by *proteins.*

Self-replicating RNA molecules appeared in the prebiotic soup due to chemical evolution and made copies of themselves using free ribonucleotides in their environment.

Lipids aggregated to form a protocell with limited ability to grow.

When the protocell replicated its genetic code, it became a true cell.

Protocell was a:

> *heterotroph* if they developed in the *primordial soup* but

> *chemoautotroph* if they were at hydrothermal vents.

Development of species from protocells

History of Earth is divided into *eons,* subdivided into *eras,* then *periods.*

Life first arose before the *Phanerozoic Eon,* the current eon.

Eons are *Precambrian Supereon,* occurring before first period of Phanerozoic, the *Cambrian.*

Precambrian encompasses more than 80% of the geologic time.

Precambrian is when the chemical and biological evolution of life, the early development of prokaryotes and eukaryotes, and the first multicellular life.

After protocell development, it evolved for millions of years until it developed into an actual unicellular organism as far back as four b.y.a.

These organisms were chemoautotrophs that resemble today's archaea prokaryotes, lacking membrane-bound organelles.

Advantageous Mechanisms

Photosynthesis emerges

Unicellular chemoautotrophs from 3.5 b.y.a. are the *last universal common ancestor* (LUCA).

LUCA diverged into domains *Bacteria* and *Archaea*. *Photosynthesis* emerged from cyanobacteria around 3.4 b.y.a. using water as a reducing agent and oxygen as a by-product.

Stromatolites, fossilized sediment excreted by cyanobacteria, are evidence of their age.

From cyanobacteria, oxygen levels in the primitive oceans rose so high over the next billion years that they converted the *reducing* atmosphere into an *oxidizing* atmosphere.

Great Oxygenation Event decimated many obligate anaerobes, which survive only in low or nonexistent-oxygen areas.

Since the Great Oxygenation Event, aerobic organisms have dominated Earth.

Endosymbiotic theory

Eukaryotic cells appeared later in the Precambrian, around 1.6 to 2.1 b.y.a.

Their exact origin is unclear. Some scientists propose that membrane-bound organelles arose from invaginations of a prokaryote's cell membrane.

Symbiogenesis proposes that two prokaryotes fuse or engulf each other by phagocytosis.

These may have been two bacteria or a bacterium and an archaeon. They developed a symbiotic relationship, with the host cell becoming a eukaryote and the other evolving into an organelle.

Endosymbiotic theory is when an anaerobic host cell engulfs an aerobic bacterium that shares its genome with the host cell and develops into a mitochondrion.

Host cells taking in a cyanobacterium developed into a chloroplast.

Mitochondria and chloroplasts

Mitochondria are in eukaryotic cells, while *chloroplasts* are only in plants and algae, suggesting that mitochondria arose before chloroplasts in evolutionary time since they are universal.

Endosymbiosis is observed in modern unicellular organisms, crediting the theory.

Some eukaryotes developed cilia and flagella to facilitate movement.

Motility was a considerable advantage in Precambrian Eon, allowing species to explore niches.

Organisms continually diversify to occupy and adapt to previously sterile environments.

Sexual reproduction

Early prokaryotes and eukaryotes were exclusively asexual, but around 1.2 b.y.a. some evolved meiosis and sexual reproduction.

Eukaryotic microorganisms that emerged are protists (e.g., algae, plankton, amoebas).

Eukaryotic microorganisms were large unicellular organisms or multicellular colonies.

It is unknown when multicellular organisms appeared, but they would have been microscopic.

Microscopic size must have been advantageous, allowing them to out-compete other organisms.

All organisms existed in the sea because the harsh UV radiation made life improbable on land.

Oxygen content of the air continued to increase until 600 million years ago (m.y.a.), when it allowed for the formation of an ozone layer in the upper atmosphere.

This ozone layer shielded most UV radiation from reaching Earth's surface, allowing organisms to live on land for the first time.

Multicellular life emerges

Near the end of the Precambrian, complex multicellular life began.

Most data come from fossils of the Ediacaran Hills in South Australia, which date to 600 m.y.a.

Organisms were soft-bodied primitive invertebrates inhabiting oceans and shallow mudflats.

Following the Precambrian came the Phanerozoic Eon.

First era of the Phanerozoic is the *Paleozoic Era* (*"ancient life"*), which includes the Cambrian Period along with five others:

> Ordovician
>
> Silurian
>
> Devonian
>
> Carboniferous
>
> Permian

Modern Organisms

Mass extinctions

Paleozoic Era lasted over 300 million years and was an active period with three major *mass extinctions*, in which many organisms went extinct over a short period.

Tectonic, oceanic, and climatic changes are possible triggers of mass extinction.

Changes may be drastic (e.g., asteroid impact) or gradual (e.g., sea-level change and global heating or cooling).

Evolution of modern phyla

Cambrian Explosion, about 542 m.y.a., marked the beginning of the Cambrian Period.

This included widespread evolution and diversification of existing prokaryotes and eukaryotes, along with the arrival of fungi and invertebrate animals.

Nearly all modern phyla evolved at this time.

Animals descended from flagellated colonial protists, with closest relatives the modern choanoflagellates.

Sponges

Colonies of the choanoflagellate ancestor perhaps gave rise to the first animals: the sponges.

Sponges could grow to one meter across, making them massive compared to existing species.

They have no nervous system and are asymmetrical.

Cnidarians and ctenophores

After sponges, *cnidarians* (e.g., jellyfish, sea anemones) and *ctenophores* (box jellies) evolved.

Cnidarians and ctenophores are symmetrical in all planes, such as *radial symmetry.*

Cnidarians and ctenophores have *rudimentary sensory organs.*

Animals higher than sponges (cnidarians, ctenophores) are *bilaterians* with bilateral symmetry.

First bilaterians were protostomes, which evolved throughout the Cambrian Period.

Protostomes and deuterostomes

Protostomes are defined by *gastrulation during embryonic development.*

They have the first opening formed in the blastula, the blastopore, which becomes the mouth.

Early protostomes were tiny, primitive, parasitic organisms, such as flatworms and nematodes.

Soon, mollusks overtook the Cambrian seas.

Marine arthropods (*crustaceans*) and early deuterostomes such as *echinoderms* (sea stars and kin) were evolving.

Deuterostomes have a blastopore which becomes an anus, with the mouth forming afterward.

Plants were simple, unicellular, and aquatic in the early Cambrian Period.

Marine algae

Marine algae expanded to freshwater throughout the Cambrian and the ensuing *Ordovician Period,* 490 m.y.a., marine algae expanded to freshwater.

Fungi began symbiotic relationships with plant roots, allowing plants to colonize bare rocks.

Invertebrate fish diverged from deuterostomes in oceans, but mollusks and crustaceans protostomes dominated.

At the period's end, 450 m.y.a., the *Ordovician-Silurian extinction* eliminated 70% of marine life, permitting fish to dominate later.

Marine invertebrates re-diversified, and crustaceans overtook mollusks as the rulers of the seas.

Terrestrial plants

Silurian Period 450 m.y.a. includes the establishment of terrestrial plants.

Silurian included the first terrestrial arthropods, the arachnids.

Advanced adaptations such as vascular structures, stems, leaves, roots, and seeds evolved over 100 million years.

Jawless fish evolved into jawed fishes, 395 to 420 m.y.a.

Jawed fish became "*armored*" with heavy, bony plates; these armored fish are all but extinct.

The armored fishes diverged into bony and cartilaginous fish in the Silurian Period.

As competition increased and available habitats decreased, cartilaginous fish adapted to be large, aggressive, and predatory, leading to shark species.

Other cartilaginous fish include skates and rays.

Age of fishes

Ensuing *Devonian Period,* 420 m.y.a., bony fish diversified into *ray-finned* and *lobe-finned.*

This period is the *Age of Fishes*, as they dominated the seas.

However, arthropods were also undergoing significant diversification, and an explosion of plant life occurred, resulting in the first true trees and seeds.

Near the end of the Devonian, as early as 395 m.y.a., *lobe-finned fishes* ventured onto land. *Lobe-finned fishes became amphibians*, the first *tetrapods (four-limbed vertebrates)*.

Late Devonian extinction 360 m.y.a. eliminated 70% of species worldwide.

Late Devonian extinction marked the beginning of the *Carboniferous Period,* which spawned the growth of forests and an abundance of seedless vascular plants such as horsetails and ferns.

Ancient plants provided carbon (*coal*).

Insects, which ventured onto land, developed wings, and exploded in diversity.

Age of amphibians

Amphibians dominated the land in the *Carboniferous Period,* marking the *Age of Amphibians*.

Some amphibians became well-adapted to land, evolving into reptiles at the end of the Carboniferous, around 300 m.y.a.

Reptiles quickly colonized land and specialized further in their terrestrial environment.

Terrestrial prey was scarce because not many animals could survive on land.

As a result, many reptiles were herbivores to take advantage of the plants on land and shorelines.

The first reptiles and amphibians to reach land and die would degrade into organic compounds.

Enriched nutrients in soil allowed plants to grow and microorganisms to exist on a larger scale.

Organisms that relied on these microorganisms migrated to land, followed by predators.

Migration to land

Migration to land continued ecological succession would permit land to support life on a scale that approached that of the sea.

After the Carboniferous came the *Permian Period*, the final period of the Paleozoic Era, when reptiles diverged into sauropsids and synapsids at the Carboniferous and Permian border.

Sauropsids were ancestors of dinosaurs, modern reptiles, and birds.

Synapsids were proto mammals resembling reptiles more than mammals:

para-mammals

proto mammals

mammal-like reptiles

During Permian, sauropsids and synapsids continued to flourish in a climate that was becoming increasingly dry; a massive desert developed in the interior of the supercontinent Pangea.

Early conifers and seeded ferns largely replaced the swampy forests of the Carboniferous.

Evolutionary Time Scales

Permian-Triassic extinction

End of the Permian, 252 m.y.a., was the greatest extinction, *Permian-Triassic extinction.*

This decimated over 95% of marine life and 70% of terrestrial vertebrates and many insects.

Like most other extinctions, the cause of this event is unknown; however, excess carbon dioxide from changing ocean circulation may have contributed to this extinction.

Permian-Triassic extinction marks the beginning of the *Mesozoic Era.*

Throughout the Mesozoic, gymnosperms flourished, but flowering plants had yet to evolve.

Marine animals diversified to re-occupy niches left by extinction.

Meanwhile, surviving reptiles became larger and evolved into dinosaurs and crocodilians.

Age of Dinosaurs characterized this era.

Triassic and Jurassic Periods

First period of the Mesozoic, the *Triassic Period,* included the evolution of dinosaurs into *avian* (or *bird-like*) and *reptilian* species.

Pangea (hypothetical supercontinent including all land masses) split into continents.

Surviving synapsids evolved into the first true mammals but were small and resembled rodents.

Triassic-Jurassic Extinction at the end of Triassic (~ 200 m.y.a.) eliminated most synapsids and almost all large amphibians; dinosaurs dominated in the succeeding Jurassic Period.

The diversity of dinosaurs was astonishing; they occupied many ecological niches.

Dinosaurs ruled the land, sea, and air, growing to mammoth proportions during the Jurassic.

Many dinosaurs were enormous compared to previous organisms. Their sheer bulk and power had an apparent selective advantage.

Dinosaurs were the most advanced tetrapods, and some moved toward bipedal motion.

These herbivores were tall and grazed on top of trees, which other organisms could not reach.

Predators, such as the Tyrannosaurus rex, the largest carnivore, became bigger and stronger.

Jurassic Period had flying reptiles such as *pterosaurs*, the famous of which is the *pterodactyl.*

Cretaceous Period

Jurassic was succeeded by the *Cretaceous Period* some 145 m.y.a.

Avian dinosaurs evolved into true birds, which dominated the skies and began to outcompete flying dinosaurs.

First flowers emerged.

Meanwhile, the continents from *Pangea* had come to resemble today's continents.

This geological change was a significant factor that molded the evolution of species.

For example, the collision of the supercontinent Eurasia with Africa allowed species from each region to interbreed, compete, or predate, greatly accelerating evolution.

Mammals emerge

From Triassic Period until the dinosaurs' last moment of glory in the Cretaceous Period, a competitor was quietly emerging: the mammals.

Although dinosaurs dominated, mammals became competitors for land resources and preyed on smaller dinosaurs.

Throughout the Mesozoic Era, mammals gradually shed their reptilian features in favor of hair, live birth (or *viviparity*), and other mammalian characteristics.

By the Cretaceous Period, mammals diverged into *placentals* and *marsupials*.

Dinosaur extinction

Causes for dinosaur extinction have been proposed.

The consensus is that a major geological event, the K-T or K-Pg extinction, extinguished dinosaurs except for avian dinosaurs, which developed into birds.

It spared many of the non-dinosaur reptiles and mammals.

Their disappearance at end of Cretaceous, 65 m.y.a., made many ecological niches available.

An asteroid may have struck the Earth, resulting in catastrophic changes to Earth's climate.

Mammals may have survived because they were smaller and required fewer resources.

Furthermore, many lived underground and could have survived an event like an asteroid.

Regardless, mammals flourished once relieved of competition and predation by dinosaurs.

Cenozoic Era

Age of mammals

K-T extinction 65 m.y.a. marks the end of the Mesozoic Era and the beginning of the *Cenozoic Era*, which extends to the present day.

It is rife with the evolution and proliferation of mammals and drastic climate changes.

Age of Mammals characterizes the entire era.

Birds dominated in the first period, the *Paleogene Period.*

Many ancient birds were large and aggressive and posed a great threat to mammals in the early years of the Cenozoic.

Primates adapted to flowering trees

In the Paleogene Period, mammals were small and inhabited overcrowded jungle environments.

Flowering plants were diverse and plentiful by the Cenozoic era, and *primates* had adapted to living in flowering trees.

First primates were small squirrel-like animals, evolving into monkeys and apes.

Early whales were semi-terrestrial but returned to the oceans later as their limbs regressed.

Later in the Paleogene, global cooling reduced jungles and the proliferation of grasslands, allowing mammals to grow large.

This trend continued throughout the *Neogene Period,* which began about 23 m.y.a.

Many modern mammals evolved, including apes.

Hominids emerge

At the end of the Neogene, some great apes evolved into *hominids,* the human-like ancestors of modern *Homo sapiens sapiens* (i.e., subspecies of *Homo sapiens*).

These apes would have moved from the jungles and ventured into the savanna.

Hominids developed bipedal motion, a significant advantage that allowed them to look across tall savanna grasses for predators and prey alike.

Bipedalism freed hands for efficient terrestrial travel in savannahs and scattered forests.

Hominids are *australopithecines,* which translates as *southern apes.*

Chronology of Earth's events

Present-day	Eon	Era	Period	Events
2.6 m.y.a.	Phanerozoic	Cenozoic	Quaternary	Homo sapiens appears Modern mammals continue to evolve Ice ages
23 m.y.a.			Neogene	Continents drift to current positions First australopithecines Modern mammals evolve and dominate Grasslands proliferate
65 m.y.a.			Paleogene	Mammals diversify and grow Birds dominate land and skies
145 m.y.a.		Mesozoic	Cretaceous	First flowering plants First true birds Mammals evolve and diversify Dinosaurs continue to dominate
200 m.y.a.			Jurassic	Dinosaurs grow, diversify, and dominate the land Mammals evolve and diversify
252 m.y.a.			Triassic	Pangea separates First true mammals First true dinosaurs Life recovers from Permian-Triassic extinction
300 m.y.a.		Paleozoic	Permian	Pangea forms Sauropsids and synapsids proliferate
360 m.y.a.			Carboniferous	Trees proliferate Reptiles diverge into sauropsids and synapsids First reptiles (evolution of amniotic eggs)

420 m.y.a.			Devonian	First amphibians Global heating First trees and seeds Arthropods diversify Bony fish diversify into ray-finned and lobe-finned fish
445 m.y.a.			Silurian	Vascular plants colonize land Arthropods colonize land Fish evolve into jawed fish, then cartilaginous and bony fish
490 m.y.a.			Ordovician	Glaciation First fish Invertebrate marine life proliferates
540 m.y.a.			Cambrian	Cambrian explosion
2.5 b.y.a.	Proterozoic			First fungi and invertebrates Ozone layer forms First protozoa First multicellular organisms First eukaryotes
4 b.y.a.	Archaean			Cyanobacteria evolve and begin producing oxygen; Great Oxygenation Event (GEO) LUCA diverges into Bacteria and Archaea First life
4.5 b.y.a.	Hadean			Cooling of crust Water condenses into oceans Reducing atmosphere Volcanic activity Celestial bombardment Earth forms

Notes for active learning

Human Evolution

Origin of humans

Scientific thought held Europe as the birthplace of hominids due to the Eurocentric outlook and the discovery of fossils in Europe.

However, throughout the 20th century, archaeologists uncovered older australopithecine remains in Africa, with Africa acknowledged as the origin of humans.

Famous of these discoveries was *Lucy* in Ethiopia in the 1980s.

Lucy represents Australopithecus afarensis, meaning *southern ape from afar.*

Southern apes from afar roamed Earth 3 to 5 m.y.a.

A. afarensis is a direct ancestor to modern humans and later species of Australopithecus.

Paleoanthropologists who discovered Lucy and other fossilized remains used skeletal features to map the subtle evolutionary changes in hominids.

The pelvis and spine determine if an organism is *bipedal* and upright or walks on all fours.

Skull and brain developments

Lucy would have been bipedal, with *stout stature* and a *small cranium.*

The skull is one of the most critical skeletal features in hominid fossils.

Large brain development and a larger skull are prominent markers of human evolution.

DNA sequencing, genomics, and molecular biology help researchers map evolutionary relationships between humans and great apes.

Hominid fossils

Australopithecus africanus superseded *A. afarensis* 2 to 3 m.y.a. from critical adaptations.

Notably, *Australopithecus africanus* had a larger cranium with more human-like facial features.

Australopithecus africanus was taller and slimmer, and their hands better adapted for tool use.

Africanus evolved into *Australopithecus robustus* and *Australopithecus boisei,* once thought to be a direct ancestor of man.

However, they became extinct, and other *A. africanus* evolved into the first genus *Homo.*

Homo habilis

Homo habilis, dated to 2.8 m.y.a. marks transition from genus *Australopithecus* to *Homo.*

Hominids exhibited intelligence like modern humans, notably by using basic stone tools.

Quaternary Period, which continues today, began 2.6 m.y.a. Early in the *Quaternary Period,* global cooling produced extensive glaciation, known as the *Ice Ages.*

Ice Ages spurred the evolution of mammals with a significant capacity to retain heat, such as giant ground sloths, beavers, wolves, bison, woolly rhinoceroses, mastodons, and mammoths.

Many species are now extinct, likely due to human hunting and environmental pressures.

While the other latitudes iced over, deserts developed in the tropics.

Homo erectus

H. habilis diverged into several groups, which are still debated.

Prevailing view is that *Homo erectus* evolved from *H. habilis* around 1.3 to 1.8 m.y.a. and migrated from Africa into Asia and Europe.

Earliest evidence of true humans has been dated to around 1.6 m.y.a.

Archaic humans included *Homo neanderthalensis* and *Homo heidelbergensis,* now extinct.

Whether these groups were subspecies of *Homo sapiens,* or a distinct species is debated.

Little consensus and clear demarcation between *Homo species* and *subspecies* are impossible.

Archaic humans spread throughout Africa, Asia, and Europe over 1 million years.

They developed wooden tools, like spears, and had cranial capacity like modern humans.

Cognitive abilities allowed them to survive in hostile environments and dominate Earth.

Humans learned to *control their environment* and *protect against natural selection pressures.*

Neanderthals

Homo sapiens sapiens is the subspecies representing modern humans, originating back 500,000 years, and persisting.

Intentional duplication of sapiens (*Homo sapiens sapiens*) distinguishes them from direct ancestor *Homo sapiens idaltu.*

Neanderthals have common ancestry and co-existence with early man.

Neanderthals became widespread across Europe and Asia around 250,000 years ago.

Once a separate genus, some accept Neanderthals as a species or subspecies of genus *Homo.*

Neanderthals' robust and stocky nature made them well-suited to cold, but *Homo sapiens* dominated deserts.

Eventually, *Homo sapiens* prevailed even in the tundra.

Denisovans

Denisovans are another early human population who lived in Asia during the Lower and Middle Paleolithic. Relatively little is known about them because archeologists have uncovered fewer fossils of these hominins.

The first discovery of a Denisovan occurred in 2010, based on mitochondrial DNA from a female bone obtained in 2008 from the Denisova Cave in the Altai Mountains of Siberia. Nuclear DNA indicates their close affinity with Neanderthals.

Archeological research suggests that modern humans overlapped with Neanderthals and Denisovans for a period, and that they interbred. Many people living today have a small portion of genes from Neanderthals and Denisovans.

Denisovans interbred with modern humans, with approximately 5% occurring in Melanesian, Aboriginal Australian, and Filipino Negrito populations. This distribution suggests that there were Denisovan populations across Asia, the Philippines, New Guinea and Australia.

Introgression into modern humans may have occurred as recently as 30,000 years ago in New Guinea. If this is correct, this population might have persisted as late as 14,500 years ago.

There is genetic evidence of Denisovans interbreeding with the Altai Neanderthals, with about 17% of the Denisovan genome from Denisova Cave deriving from them.

Age of man

Opinion divides on whether modern humans outcompeted the Neanderthals or incorporated them via interbreeding.

Neanderthals vanished, and *Homo sapiens sapiens* were the sole humans 40,000 years ago.

Homo sapiens sapiens eventually inhabited all continents except Antarctica.

Homo sapiens characterizes the Quaternary Period as the "Age of Man."

Many assert that a sixth mass extinction, driven by humans, signals the end of the Quaternary within the unknown future.

Notes for active learning

Notes for active learning

REVIEW

Classification & Diversity

Page intentionally left blank

Common Ancestry: Shared Conserved Features

Common descent theory

Common descent theory is a central tenet of evolutionary biology. This theory states that any given group of organisms share a common ancestor from which they descended.

It naturally follows that there is a common ancestor to every organism on Earth, called the *last universal common ancestor* (LUCA).

Charles Darwin (1809-1882) and his contemporaries gathered evidence about the foundation of this theory, but modern advances in molecular biology have provided many new insights.

In contemporary biology, life is divided into three domains:

Archaea,

Bacteria, and

Eukarya

Archaea constitute a separate domain from Bacteria in recent years.

Bacteria and archaea are unicellular prokaryotes but are not classified in the same domain.

LUCA first diverged into bacteria and archaea, with eukaryotes developing from these.

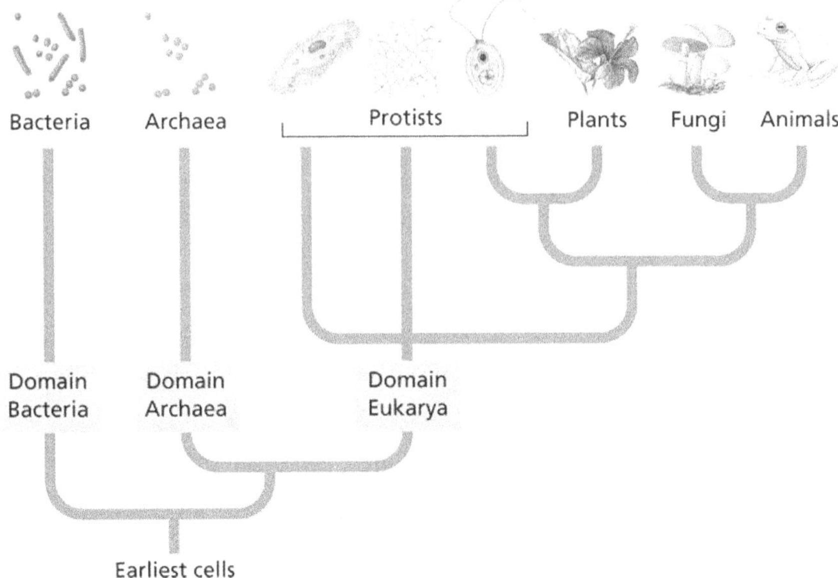

Three domains of Bacteria, Archaea, and Eukarya with proposed kingdom relationships

Structural and functional relatedness of domains

Archaea live in a wide range of habitats, but the most well-known species are the extremophiles, which inhabit hostile environments like hot springs, salt lakes, and anaerobic swamps.

Bacteria and archaea have prokaryotic features, such as a cell wall (archaea lack peptidoglycan) but are different biochemically.

Domain Eukarya is a diverse group of unicellular and multicellular organisms, the most well-studied, and includes kingdoms Animalia, Plantae, Fungi, and Protista.

Evidence from fossils, comparative anatomy, and experimentation began in the 18th century.

Modern evidence is DNA and RNA sequences, analysis of common enzymes and metabolic pathways, fundamental similarities in cell structure, and ongoing evolution.

All three domains have unique differences, but their underlying structure and function display undeniable evolutionary similarities.

Biochemical relationships

Biochemical evidence is some of the most compelling.

Life forms use biomolecules similarly.

Organisms use fundamental building blocks, genetic codes, and metabolic pathways.

Specific genes are largely conserved across all domains of life, coding for proteins that can be remarkably similar or identical.

For example, organisms use DNA and RNA to carry genetic information encoded in base nucleotides (A, T, G, C, and U).

All organisms interpret nucleotide bases as triplet codes (i.e., *codons*) to synthesize proteins from twenty amino acids, supporting the hypothesis that all three domains descended from a common ancestor.

Furthermore, the code and method of genetic interpretation are widespread.

Many organisms share introns (noncoding mRNA sequences), which is notable because these are inoperative, and it is unknown why they are so similar.

Structural evidence of eukaryotic relatedness

Eukaryotes are the most well-researched domain; they are readily observable and of the most interest to humans.

Consequently, there is a wide array of evidence for the relatedness of all eukaryotes.

Eukaryotic structural features include an extensive cytoskeleton and organelles such as mitochondrion, chloroplast, and nucleus.

These organelles are part of the endomembrane system, a collective term for membranous sacs dividing a cell into its constituent parts.

In contrast to bacteria and archaea, eukaryotes are the only organisms with this system of membrane-bound organelles.

Similarities among eukaryotes include DNA-bound linear chromosomes and shared metabolic pathways, such as glycolysis.

The earliest evidence for the structural relatedness of eukaryotes comes from fossils.

Fossils may be skeletons, shells, seeds, imprints, and even soft tissues.

Fossil record is the history of life recorded by past remains.

Unfortunately, fossil records are often incomplete because most organisms decay before fossilization, and soft-bodied organisms generally do not fossilize.

Relative dating

Sedimentation began when Earth formed; particles accumulated, forming a stratum as a recognizable layer in a sequence of strata.

Relative dating places it in the sequence indicating the age of a fossil; a stratum is older than the one above and younger than the one below.

Most fossils are embedded in or recently eroded from sedimentary rock.

Relative dating does not establish the absolute age of fossils.

Radioactive dating

Radioactive dating establishes the absolute age of fossils.

Radioactive dating uses radioactive isotopes with a quantifiable half-life.

Half-life is the time it takes for half of a radioactive isotope to decay into a stable element.

Carbon-14 (^{14}C) is a radioactive isotope contained within organic matter.

Half of carbon-14 decays into nitrogen-14 every 5,730 years.

Comparing ^{14}C radioactivity of a fossil to modern organic matter calculates the fossil's age.

However, after 50,000 years, carbon-14 radioactivity is too low to measure age accurately.

Paleontologists use potassium-40 and uranium-238, which have half-lives of billions of years.

Fossil record

Paleontologists use *strata* (*flat layers of sedimentary rock*) and *fossils* (*ossified* or *petrified* organisms) to study the history of life.

An essential principle of fossil study (i.e., paleontology) is that an organism most closely resembles a recent fossil in the line of descent.

Underlying similarities become fewer farther back in the lineage.

Similar fossils can be used to construct a timeline showing changes from a common ancestor millions of years ago to a modern species.

For example, transitional forms such as bird-like dinosaur Archeopteryx establish birds descended from reptiles.

Fossils can reveal a wealth of information about an extinct animal, as in the case of the horse ancestor *Hyracotherium*, small with cusped, low-crowned molars, four toes on each front foot and three on each hindfoot.

These adaptations for forest living were gradually replaced by larger size, grinding teeth, and reducing toes into hooves as the forests gave way to grasslands.

Intermediate forms transition between *Hyracotherium* and modern horse, genus *Equus*.

However, it can be challenging to identify the proper patterns in the fossil record and ignore misleading ones. Fossils are continually reclassified and reevaluated.

For example, paleontologists have studied turtle fossils and proposed that they are closer to crocodiles than previously thought.

Biogeographical evidence

Biogeographical evidence provides a valuable frame of reference for the fossil record.

Biogeography examines the distribution of organisms across the Earth, providing additional insight into evolutionary research.

Physical factors, such as the location of continents, determine where a population can spread and influence evolution.

Forms evolve in one locale and spread to other areas, explaining the distribution of organisms.

Biogeography, as applied to evolution, is sometimes known as *phylogeography*.

For example, in the 19th century, Darwin observed that the Galapagos Islands had a variety of finch species, but the mainland had one.

Darwin concluded that the Galapagos finches had originated on the mainland and diversified once they reached the islands.

Notes for active learning

Phylogenetic Trees

Three domains of life

Systematics (or *classification*) is the study of the diversity of organisms using evidence from the molecular to the population level.

*Systematic*s is often used synonymously with *taxonomy*, although it refers to *nomenclature*, a subset of broader classification.

Phylogeny of species reveals evolutionary relationships between organisms, past and present.

Systematics may be applied to modern or extinct species included in phylogeny.

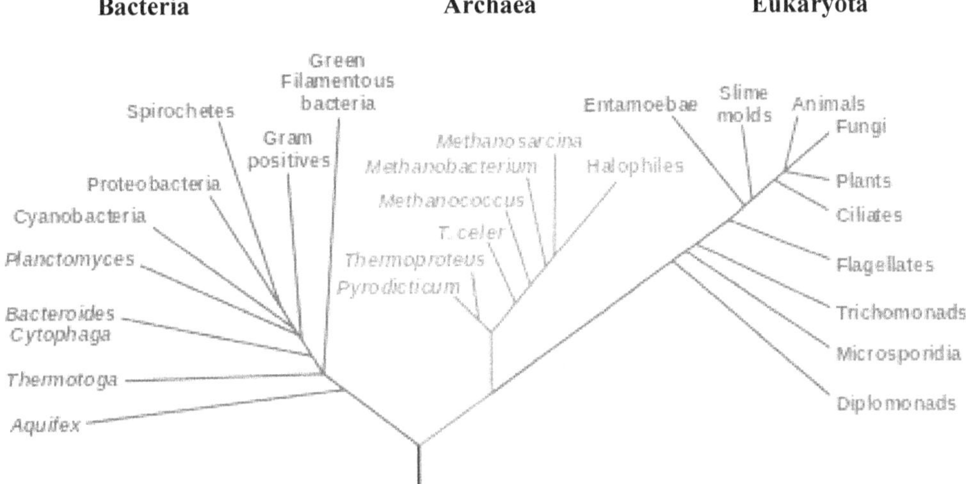

Phylogenetic tree of life with three domains, nodes, and branches

Phylogenetic descent

Phylogenetic trees visually indicate common ancestors and lines of descent.

Common ancestor is where two branches diverge at the *node*.

Branch tips are labeled with an individual taxon, usually a species.

Phylogenetic trees are represented by a branching diagram resembling a tree.

Often the tree includes an outgroup that does not share the recent ancestor of other groups on the tree. Outgroups provide a reference for the taxa of interest.

Phylogenetic trees can be constructed broadly or narrowly, from all life to a species' subforms.

Traditional phylogenetic tree branches correspond to the relative length of evolutionary time.

Pleisomorphy

Pleisomorphy (*near form*) and *symplesiomorphy* are synonyms for an ancestral character shared by clade members, which does not distinguish between clades.

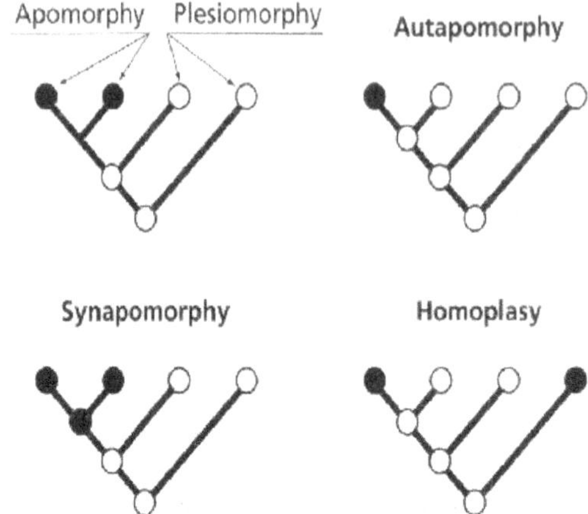

Pleisomorphy with ancestral (open circles) and derived (closed circles) traits highlighted

Traits derived or lost due to evolution

From the diversity of life, each species on a phylogenetic tree must have lost traits seen in the common ancestor and adopted new ones.

Natural selection influences these changes by the viability of traits in each environment.

In systematics, *traits* (or *characters*) are structural, chromosomal, or molecular features distinguishing taxa.

Phylogenetic trees compare which characters are shared between organisms.

For example, members of a kingdom share general characters, whereas members of a species share specific characters.

These general characters, *plesiomorphies* (primitive characters or ancestral states), are traits present in a common ancestor and descending group members.

For example, a vertebral column is a plesiomorphy of vertebrates.

Apomorphy

Apomorphy (or *derivate character*) are traits found in a few descendants.

Lineages diverge from a common ancestor with apomorphies but share *pleisomorphy*.

Apomorphies are in a *specific line of descent*.

Primitive and *derived characters* are used, but phylogeneticists prefer *plesiomorphy* or *apomorphy*.

Primitive implies that the ancestral character was rudimentary, which is not always true.

Derived implies that the character is not found in the ancestor but arose in the descendant. This is not necessarily true; a derived character can represent the absence of a trait.

For example, some reptiles became snakes as they gradually lost their limbs. The common reptilian ancestor had four limbs.

The characteristic of four limbs was a *plesiomorphy* of reptiles (and all tetrapods).

The characteristic of having lost them is an *apomorphy* shared amongst snakes.

Speciation and common ancestry

Speciation event is when species diverge from a tree node.

Defining speciation is a complex process aided by examining the species' genetic code for nucleic acid and amino acid sequence changes.

Advances in this analysis have made abundant data available to researchers.

Comparing homologous (*similar*) DNA and RNA sequences elucidate the degree of difference between organisms.

A high degree of homology indicates highly related organisms, while different sequences indicate unrelated organisms.

Biomolecular studies may show that organisms resembling each other are not closely related.

For example, the red and Chinese giant pandas have false thumbs, but the former has raccoon-like teeth and bones, while the latter does not.

DNA analysis confirms that the two pandas are distantly related.

Red pandas originated from the same lineage as *raccoons*.

Giant pandas originated from the same lineage as *bears*.

Notes for active learning

Related Species Comparisons

Molecular clocks

Rate of mutations in the DNA nucleotide sequence is relatively constant and provides a "molecular clock."

Molecular dating compares the relative age of two organisms, quantifying the number of nucleotide base pair changes between their genomes.

Comparing trace DNA from two fossils thought to be of the same lineage can determine which organism the ancestor was and how long it took the other organism to evolve from the ancestor.

Mitochondrial DNA

Mitochondrial DNA (mtDNA) is often used for molecular dating because it changes about ten times *faster* than nuclear DNA, making it helpful in *comparing related species*.

Mitochondrial DNA passes from mother to offspring, removing the effect of recombination for chromosomal DNA and elucidating the timeline.

Fossil records calibrate the clock and confirm a hypothesis drawn from molecular data.

For example, mitochondrial DNA sequences have a 5.1% nucleic acid difference between a songbird's common ancestor and descendant.

Molecular clock indicates descendants diverged from a common ancestor 2.5 million years ago.

mtDNA is less useful for distantly related organisms.

For distantly related organisms, RNA or nuclear DNA is used.

Amino acid sequences

Protein comparison determines speciation or common ancestry, albeit less robust.

Amino acid sequences of homologous proteins can be compared, as with cytochrome c, to a protein found in all aerobic organisms.

Protein analysis is limited to distant relations since few proteins are found across domains.

Biomolecular evidence is invaluable for phylogeneticists, but it is helpful with other evidence.

For example, the 140 amino acid composition of cytochrome c differs between chickens and humans by twelve amino acids (8.6%) but only three (2.1%) between chickens and ducks.

This evidence of amino acid composition makes it reasonable that humans, chickens, and ducks share a common ancestor but that chickens and ducks are more related than humans.

Evolutionary relationships to extinct species

Evolutionary relationships between living and extinct species are observable in comparing the function and development of their morphology and external and internal anatomy structures.

Living species appear morphologically like a recent ancestor, with modifications.

A distant ancestor may appear different compared to the modern organism.

However, further analysis would reveal the evolutionary link between them.

George Cuvier (1769-1832), a French vertebrate zoologist, first used comparative anatomy to classify animals.

Cuvier recognized that organisms often have similar morphological similarities because of common descent.

Homologous structures are similar anatomical traits derived from the same ancestor but modified over time.

For example, vertebrate forelimbs contain the same sets of bones organized similarly, despite their distinct functions.

A whale's flippers and a human's forelimbs are homologous structures, providing evidence of divergent evolution from a common tetrapod ancestor.

Morphological similarities

Analogous structures are inherited from unique ancestors but have come to resemble each other because they serve a similar function.

For example, fish and marine mammals have sleek, hydrodynamic body plans but evolved this trait separately.

Although the shape is present in their common ancestor, a primitive fish, it was lost in mammals and later regained by a few.

Therefore, a streamlined shape is an analogous structure that has evolved separately many times in response to an aquatic environment.

Morphology classifications can be misleading due to analogous structures.

Organisms may appear related due to their similar morphologies, but genetic or fossil analysis may reveal that these traits evolved separately.

Vestigial structures

Traits without present-day advantages are present in organisms because they do not negatively affect their fitness.

Vestigial structures are regressive and nonfunctional traits in the modern organism but serve a purpose in an ancestor.

For example, whales' vestigial hind limbs harken to their tetrapod origins.

Losing vestigial structures is an ongoing process well-studied in humans.

For example, humans no longer eat plant roughage, which requires extra molars to break down.

As a result, human jaws are becoming smaller and forcing out wisdom teeth.

In modern times, hundreds of millions of people lack wisdom teeth altogether.

Another example of a *vestigial structure* in humans is the coccyx, a bone at the end of the spine (i.e., tailbone), a regressed tail.

Notes for active learning

Clades and Cladograms

Taxonomic classifications

The first unified form of systematics arose in the 18th century when Carl Linnaeus (1707–1778) devised a taxonomic classification system.

Taxonomy assigns a hierarchy and requires each taxon to have a rank relative to other taxa.

The logical idea was that species have a fixed place in the divine hierarchy of God's plan.

However, assigning ranks to groups can be confusing and prone to bias.

Countless modifiers were added to the simple taxa of kingdom, phylum, class, order, family, genus, and species, leading to infraclass, subspecies, and superphylum variations.

Linnaeus' method did not acknowledge evolution and included extinct species.

Linnaeus' original classifications

Linnaeus' original scheme divided life into the kingdoms Animalia, Plantae, and the third kingdom for minerals (since abandoned).

In the 20th century, a five-kingdom classification was widespread, including kingdoms Animalia and Plantae and kingdom Monera (for prokaryotes), Protista (for protists), and Fungi.

Prokaryotes are divided into archaea and bacteria.

Monera was eliminated in 1983, and kingdoms of Bacteria and Archaea were established.

To represent bacteria and archaea equal to eukaryotes, they are named domains.

Each domain includes one kingdom of the same name.

Protists, animals, plants, and fungi are eukaryotes; Eukarya unified them under one domain.

It would be misleading to conclude that life consists of the domain Eukarya, kingdom Bacteria, and kingdom Archaea because eukaryotes did not come any earlier than the other two and are no longer "overarching" of a group.

This example reveals how taxonomy can become convoluted due to ranking.

Evolutionary taxonomy

Taxonomy is the naming, defining, and classifying of organisms based on shared characteristics.

Taxonomy is part of a larger whole: *phylogeny*.

Phylogeny is the evolutionary history and relationships among or within groups of organisms. However, it was too constricted, and there was no relationship consensus.

Darwinian taxonomy (or *evolutionary taxonomy*) includes species past and present and relies on common descent.

Darwinian taxonomy did not construct orderly lists but showed phylogenetic relationships.

Phenetic systematics

Phenetic systematics is a quantitative classification that arose in the 20th century.

It constructs phylogenetic trees based on statistics, counts shared characteristics between species, and analyzes the tree.

Phenetic systematics cluster species based on the number of shared characteristics.

Phenetic systematics does not distinguish between *plesiomorphies* (i.e., ancestral traits) and *apomorphies* (i.e., specialized or derived traits).

Phenetic systematics does not consider subtle evolution (e.g., parallel, divergent, convergent, and co-evolution).

This classification is vulnerable to misleading similarities, such as analogous structures or those which have evolved alongside one another.

Practitioners do not believe that a classification reflecting phylogeny can be constructed; relying on a method that eliminates nuance is better.

Phenograms depict results. Phenograms vary depending on how data is collected and processed.

Cladistics

Cladistic systematics (or *cladistics*) supersede phenetic systematics.

Cladistics requires complex mathematics for how evolution influences characteristics.

Analysis of characteristics classifies organisms into *clades*.

Cladistics is guided by the principle of *parsimony*, stating that solutions with the least assumptions are the most logical.

According to parsimony, the best cladogram has minimal assumptions, generally the simplest.

Parsimony minimizes bias but depends on the investigator's knowledge, skill, and evidence.

Like earlier methodologies, cladistics is prone to ambiguity and confusion.

It is difficult to pinpoint a common ancestor or articulate what makes two organisms evolutionarily related.

Cladistics needs a consensus on which characters should be used to distinguish groups.

Clades

Clade is any taxon with a common ancestor; it may be broad or narrow.

Clades are organized into *cladograms*, showing species relationships by shared characteristics.

Common ancestor branches are equidistant and do not correspond to relative evolutionary time.

*Cladogram*s, unlike phylogenetic trees, do not infer how much species have changed but that they have diverged.

Monophyletic clades include a given common ancestor and its descendants.

Paraphyletic clades have a *common ancestor* and many *descendants*, but not all.

Monophyletic clades are the most useful in phylogeny, but paraphyletic clades provide insight.

However, there is seldom a compelling cause to construct a *polyphyletic* clade with descendants who do not share the same common ancestor.

Polyphyletic clades are remedied into monophyletic clades.

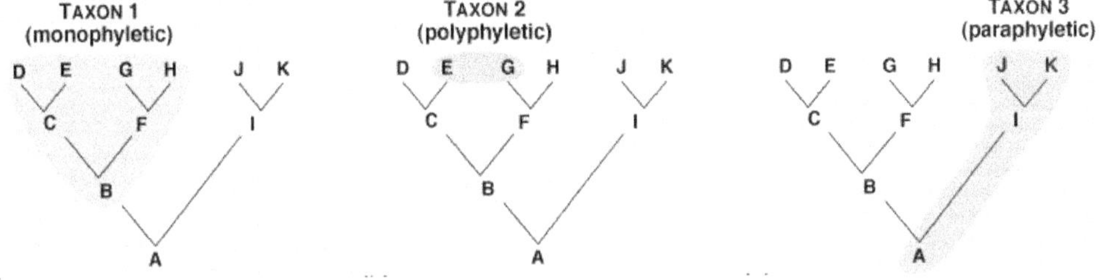

Monophyletic, polyphyletic, and paraphyletic clades and descendent relationships

Notes for active learning

Comparative Anatomy

Molecular evidence supports morphology

Sequence analysis and molecular dating are integral to modern phylogenetics, but morphology remains the simplest and most accessible way to describe the relationships between organisms.

This text uses comparative anatomy to present the diversity of plant and animal kingdoms.

Molecular evidence underpins these anatomical traits during classification.

Diversity of Kingdom Plantae

Kingdom Plantae is a clade of eukaryotes that are autotrophic by photosynthesis.

Kingdom Plantae clade usually includes multicellular green algae, terrestrial *nonvascular plants* (or *lower plants*), and terrestrial *vascular plants* (or *higher plants*).

Many phylogeneticists exclude multicellular green algae since inclusion arguably makes the clade paraphyletic; unicellular green algae are in the kingdom Protista.

Algae and *bryophytes* (*terrestrial nonvascular plants*) are green, low to the ground, and found in moist areas.

Algae and bryophytes do not have a vascular system or true stems, leaves, and roots.

Root-like *rhizoids* anchor them.

Capsules at the end of stalks protrude from *bryophytes* (gametangia and sporangia) and release spores into the wind or water.

Algae and bryophytes reproduce via these spores rather than seeds.

Vascular plants

Vascular tissue genesis allowed plants to reach unprecedented heights.

Vascular plants include seed and seedless plants.

Well-known phylum of *vascular seedless plants* is ferns, *filicophyta,* or *polypodiophyta.*

Vascular plants have true roots, leaves, and stems but reproduce using spores like nonvascular plants.

Spores are produced on the underside of long fronds divided into many leaflets.

In the Paleozoic Era, ferns ruled the forests, becoming so large that they resembled trees.

Today, most ferns are much smaller, with a maximum height of about 15 meters.

Fern allies are seedless plants classified under the phylum *Lycopodiophyta*. These plants designated as "mosses" are not mosses.

Lycopods include club mosses, spike mosses, and quillworts and are distinguished from ferns by their *microphylls,* simple leaves with a single vein.

Ferns have *megaphylls,* which have complex, branching veins.

Spermatocytes

Spermatophytes (or *seed plants*) are a monophyletic clade of plants reproducing with seeds.

Seed plants are the largest, most recent, and most successful taxon of plants.

Spermatophytes are divided into:

> *angiosperms* (or *flowering plants*) and

> *gymnosperms* (*cone plants*).

Angiosperms and gymnosperms

Angiosperms have their seeds fully enclosed within the ovaries of flowers.

Gymnosperms have their seeds partially exposed on the scale-like leaves of cones.

Both groups include the tallest plants: angiosperms with Eucalyptus, which reach 100 meters, and gymnosperms with redwood, up to 115 meters.

Angiosperms have vascular tissues, roots, stems, leaves, flowers, and fruit.

Flowers contain *ovaries* enclosing *ovules.*

When pollinated by a male flower, *ovules* are fertilized and develop into *seeds.*

Ovaries ripen around seeds to form fruit, dispersed by wind, water, gravity, or animals.

Once dispersed, the seed germinates in the soil and grows.

Gymnosperms have neither flowers nor fruit.

Gymnosperms are typically woody; conifers are the most well-known (e.g., pines and fir trees).

Gymnosperms have ovules on female cones, called *ovule cones* or *seed cones.*

Pollen from male cones (*pollen cones*) fertilizes an ovule cone and causes seed development.

Ovule cone drops to the ground for seeds to germinate.

Kingdom Animalia

Porifera: Sponges

Kingdom Animalia is a monophyletic clade of multicellular eukaryotes, heterotrophic by ingestion and motile.

Sponges (phylum *Porifera*, meaning *pore bearer*) are primitive animals that are asymmetrical and lack specialized tissues or organs.

Unlike other animals, sponges are *sessile* (or *stationary*) without a rudimentary sensory system.

Sponges are *diploblastic* with two cell layers.

Sponge intakes water through pores lined with *choanocytes*, flagellated cells that filter food and expel the remaining water through their mouth, the osculum.

Choanocytes pass food to amoebocytes for digestion and distribution of nutrients.

Many sponges have a *calcium carbonate exoskeleton* secreted by cells. They may feature *spicules*, skeletal needles of silica, or calcium carbonate that *strengthen their body walls*.

Cnidarians: Corals

Cnidarians (phylum *Cnidaria*) evolved shortly after sponges.

Cnidarians, like sponges, are *diploblastic* but exhibit complex cell specialization.

A gelatinous layer of *mesoglea* between their two cell layers develops into *muscle-like tissue*.

Cnidarians include jellyfish, hydra, sea anemones, and corals.

Cnidarians may have a sessile form, the *polyp*, or a swimming, bell-shaped form, the *Medusa*.

Many cnidarians have a larval polyp stage and a medusa adult stage. They have symmetry.

Polyps may live in colonies (coral), while medusae are free-living (jellyfish).

Cnidarians have one opening for ingesting food, expelling waste, and conducting respiration.

They have a rudimentary nervous system and sensory organs, with nothing resembling a brain.

Cnidarians are noted for their *nematocysts,* stinging cells to stun or kill prey.

Comb jellies (phylum *Ctenophora*) are diploblastic organisms with radial symmetry.

Comb jellies are grouped with cnidarians, but recent evidence suggests they may be closer to the higher-level *bilaterians*.

Bilaterians

Bilaterians are more complex animals than sponges, ctenophores, and cnidarians.

Bilaterians are triploblastic from three germ layers: ectoderm, mesoderm, and endoderm.

They exhibit bilateral symmetry and have two distinct, symmetrical sides with few exceptions.

Bilaterians exhibit *cephalization,* the concentration of nervous tissues at its anterior.

Platyhelminthes: Flatworms

Worms (phylum *Platyhelminthes*) were among the first bilaterians with phyla, including:

> flatworms of phylum *Platyhelminthes,*

> nematodes of phylum *Nematoda,* and

> segmented worms of *Annelida.*

Despite umbrella term *worms*, three phyla are more related to other organisms than each other.

Flatworms are the simplest without body cavities (*coeloms*), making them *acoelomates* (i.e., lacking a coelom).

Flatworms undergo respiration by simple diffusion and have one digestive opening.

Flatworms, like carnivorous planarians, may be:

> *free-living* or

> *parasitic* (e.g., flukes and tapeworms).

Rotifers

Rotifers (phylum *Rotifera*) are their closest relatives, microscopic filter feeders.

Rotifers are *pseudocoelomates with* a coelom, not entirely lined by mesoderm-derived tissue.

Nematodes: Roundworms

Roundworms (phylum *Nematoda*) are *pseudocoelomates.*

Roundworms inhabit various habitats and have complex respiratory and digestive systems.

Often, they are free-living soil dwellers that help decompose and recycle nutrients.

Annelids: Segmented worms or ringed worms

Annelids (phylum *Annelida*), such as earthworms and leeches, are true *coelomates.*

Annelids have complex, segmented bodies; each segment contains organs identical to those in other segments.

They move by muscle contraction, aided by bristles (*setae*) and *parapodia* limbs.

Annelids' closest relatives are mollusks.

Mollusks

Mollusks (phylum *Mollusca*) are the largest marine phylum.

Triploblastic, true coelomates with a rudimentary nervous system and some cephalization.

Usually, mollusks have a mantle, radula, and foot.

> *Radula* is a raspy tongue.

> *Foot* is used for motility, sensation, and anchoring to surfaces.

> *Mantle encloses a cavity for respiration and excretion.*

Evolution in the Cambrian seas would have given early mollusks advantages in motility, protection, and feeding.

Many classes are in the phylum Mollusca, with three widely known.

Bivalves, gastropods and cephalopods

Bivalves (class *Bivalvia*), such as clams and oysters, secrete a hinged shell from the mantle.

They are generally filter feeders on ocean floors, tidal areas, or even free swimming.

Gastropods (class *Gastropoda*), the snails and slugs, have a single continuous shell and are noted for their large, developed feet.

Most gastropods are herbivorous grazers.

Cephalopods (class *Cephalopoda*) are the most advanced mollusks; these include octopus, squid, cuttlefish, and nautilus.

Cephalopods have well-developed cephalization and are primarily predatory, using tentacles to attack prey.

Arthropods

Arthropods (phylum *Arthropoda*) include an incredibly diverse phylum on land, sea, and air.

Jointed limbs, segmented body plans, and exoskeletons at a basic level characterize arthropods.

Exoskeleton is *chitin,* a polysaccharide that undergoes continuous molting and replacement.

Small coelom surrounds reproductive and excretory systems, with a body cavity called a *hemocoel* facilitating an open circulatory system.

Coelomic fluid and blood flow through the hemocoel, bathing surrounding tissues.

Arthropods have an advanced nervous system with a primitive brain and sophisticated photoreceptors, mechanoreceptors, and chemoreceptors.

Metamorphosis

Arthropods exhibit one of two life cycles when an immature:

> *nymph* gradually develops into an adult, or

> *larva* enters a cocoon and undergoes metamorphosis (sudden development) into adulthood.

Phylum *Arthropoda* divides into four subphyla.

Trilobita, a now-extinct fifth subphylum, comprised trilobites.

Arthropods were ubiquitous in the Cambrian and later periods.

Arthropods: Crustaceans

Crustaceans (subphylum *Crustacea* or *insects of the sea*) are arthropods.

Crustaceans include lobsters, crabs, shrimp, krill, and barnacles.

Many crustaceans are herbivores and of immense importance to aquatic ecosystems.

They feed on phytoplankton, making them a crucial link in food chains.

Few crustaceans colonized land and freshwater environments, but most are marine, using gas exchange by gills.

By contrast, spiders (subphylum *Myriapoda*) are exclusively terrestrial.

Arthropods: Spiders

Spiders are arthropods in the subphylum *Myriapods*, including *centipedes* and *millipedes*, which use many feet.

Centipedes have fewer legs than millipedes and are fast, effective predators with venomous glands which assist in killing prey.

Millipedes are slow, so they are primarily herbivores and detritivores.

Millipedes have venomous glands, which they use as chemical defenses against predators.

Along with class *myriapods*, *Arachnida* (subphylum *Chelicerata*) were among the first creatures to occupy the land.

A few arachnids have returned to the water.

Famous arachnids are spiders, scorpions, ticks, and mites.

Arachnids have eight legs and advanced eyes, making them excellent predators.

They exhibit unique "book lungs," stacks of alternating air pockets for gas exchange.

Arthropods: Insects

Insects (class *Insecta*, subphylum *Hexapoda*) are Earth's *most successful and diverse taxon.*

There are more insect species than other species combined.

Insects are characterized by:

> three-part body plans,

> six legs, and

> antennae but have developed additional features.

Wings allowed insects to escape from predators and travel large distances.

Primitive insects (likely the first) are wingless, with flying as an evolutionary adaptation.

Most insects inhabit land and air.

Respiratory systems have *spiracles* (openings), allowing air to enter internal *tracheal tubes.*

Insects are limited in size because their anatomy *cannot support a large body.*

Most insect species are small, occupying small areas and minimizing their food requirements.

Triploblasts

Triploblasts (Platyhelminthes, nematodes, annelids, rotifers, mollusks, arthropods, and other minor phyla and classes) are collectively classified as *protostomes*.

Taxa are debated; only echinoderms and chordates are definitively deuterostomes.

Researchers disagree on whether *bryozoans*, a phylum of marine invertebrates, are protostomes or deuterostomes.

Brachiopods, bottom-dwelling animals resembling mollusks, are a phylum unresolved.

Brachiopods have been classified as *deuterostomes*, but new evidence suggests they may be *protostomes*.

Echinoderms: Sea urchins

Echinoderms include sea stars, sea urchins, cucumbers, sand dollars, brittle stars, and sea lilies.

Echinoderms are *bilaterians,* like all organisms higher than cnidarians and ctenophores.

Which deuterostome evolved first is inconclusive; evidence suggests the phylum *Echinodermata.*

Echinoderms evolved radial symmetry after descending from bilateral ancestors (bilateral larvae).

Echinoderms had strong exoskeletons and accomplished camouflaging abilities.

Echinoderms do move.

Some are herbivores, but many sea stars are carnivorous and move surprisingly quickly.

Complex coelom

Echinoderms have robust regeneration powers and complex coelom, including a water vascular system for water pressure to extend *tube feet*, allowing the animal to move.

Tube feet also function in the respiratory, digestive, and nervous systems.

Because nearly all chordates are vertebrates, the two taxa are often conflated.

Advanced deuterostomes belong to the phylum Chordata and are primarily defined by developing a notochord in the early embryo.

Notochord develops into a vertebral column, so they are classified in the subphylum Vertebrata.

Non-vertebral chordates include tunicates and lancelets, simple filter feeders, and likely the earliest chordates.

Chordates: Vertebrates

Notochord

Chordates possess a *notochord* during embryogenesis.

Notochord is a flexible rod below the dorsal nerve cord along the length of chordate embryos.

Notochord is derived from *mesoderm* (i.e., middle germ layer) and develops in *neurulation* during embryogenesis.

In non-vertebral chordates, the notochord persists after birth; muscles work against it to help the animal move.

The notochord may be lost in sexual maturity or remain throughout adulthood.

Notochord is replaced by *vertebral column* for vertebral chordates in embryonic development.

Pharyngeal pouches and branchial arches

Pharyngeal pouches, present in chordate embryos, are a distinguishing feature of chordates.

Pharyngeal pouches are openings between the pharynx (a portion of the throat) and the environment. They develop into various structures in organisms.

Non-vertebrate chordates have pouches that develop into *pharyngeal slits* for filter feeding and gas exchange.

In fish and some amphibians, the pouches are modified into the jaw, the hyoid bone, and the *branchial arches*.

Branchial arches are *gill arches* because they support the gills.

In some vertebrates, the pouches form *pharyngeal arches,* which develop into the jaw, hyoid bone, thyroid, larynx, tonsils, and other components of the neck and head.

Dorsal nerve cord

Dorsal nerve cord (derived from ectoderm) forms the *central nervous system*, a hollow chord extending the length of the embryo on top of the notochord.

Like the notochord, the dorsal nerve cord develops during *neurulation.*

Most dorsal nerve cord develops into the spinal cord in vertebrates, and the anterior part enlarges into the brain.

Post-anal tail extends at the posterior end and may be lost or retained in the adult organism.

Vertebrate phylogeny

The first vertebrates likely were jawless fishes like *hagfish* and *lampreys*.

Primitive filter feeders diversified greatly after the Cambrian, some evolving into jawed fishes.

Jawed fish evolved into a cartilaginous jaw and skeleton; today, they are cartilaginous fishes (class Chondrichthyes).

Notochord is replaced by cartilage rather than bone in this class (e.g., sharks, skates, and rays).

Cartilaginous fishes have tough skin and are typically predatory.

Bony fish have a bony jaw and skeleton and are divided into:

>> *ray-finned fish* (class *Actinopterygii*)

>> *lobe-finned fish* (class *Sarcopterygii*)

After cartilaginous fish, bony fish (superclass *Osteichthyes*) split from their common ancestor.

Bony fish are the origin of nearly all fish species.

Lobe-finned fish

Lobe-finned fish are primitive; few survived, while ray-finned fish represent most fish species.

Most fish are *ectothermic* (*cold-blooded*); body temperature depends on the environment and cannot be regulated.

Most breathe with gills, pumping blood past gills with *two-chambered hearts*; a few fish have rudimentary lungs.

Fish diverge significantly in morphology but typically are streamlined to optimize swimming.

Typical characteristics of fish include *scaly skin*, *fins*, and reproduction with *eggs*.

Tetrapods

Ancient lobe-finned fish that developed lungs ventured onto land using their fleshy, lobed fins.

These are ancestors of lungfish and all terrestrial vertebrates, known as *tetrapods* (superclass *Tetrapoda*) for four limbs.

This group is class Amphibia, Reptilia, Aves, and Mammalia.

Tetrapods were the origin of animals such as snakes and marine mammals.

Many tetrapods lost most limbs over time, and some returned to the water.

Amphibians

The earliest tetrapods, the amphibians (class Amphibia), never fully left their aquatic origins.

Most amphibians are adapted for terrestrial and aquatic habitats, and many spend the entire first stage of their lives in water.

Larval forms typically breathe with gills but undergo metamorphosis into adults, breathing with lungs and skin.

Skin is moist and can exchange gases and water with the environment.

Two-chambered hearts pump blood throughout their closed circulatory system.

Like fish, amphibians are *ectotherms*.

Three main orders of amphibians are:

> *frogs and toads* (order *Anura*)
>
> *salamanders* (order *Caudata*)
>
> *caecilians* (order *Gymnophiona*)

Caecilians are primitive, limbless amphibians that resemble giant worms.

As ancient amphibians thrived, an important apomorphy arose.

Amniotic eggs

Amniotic egg is a specialized egg with complex membranes that can sustain an embryo on land or, in the case of mammals, within the womb.

Embryos fully develop within eggs before hatching, unlike fish and amphibian embryos in the larval stage when they exit the egg.

Earlier organisms, such as amphibians and fish, laid eggs in water.

These eggs are simple and gelatinous and exchange materials with their aquatic environment.

By contrast, amniotic eggs can be laid on land and exchange gas with the air while protecting the embryo with its hard, leathery shells.

Reptiles

Amphibians who evolved amniotic eggs and scaly, waterproof skin adaptations became reptiles.

Reptiles quickly colonized land and specialized further in their terrestrial environment.

Reptiles are widespread and diverse.

Turtles, tortoises, and terrapins (order *Testudines*) inhabit the seas and land, protected by their large, tough shells.

Lizards and snakes (order *Squamata*) are the most diverse reptile group, with nearly 10,000 known species.

By contrast, crocodilians (order *Crocodilia*) have 25 living species, and the tuatara (order *Sphenodontia*) is the only reptile of its kind.

Many reptiles can maintain a remarkably stable body temperature despite being ectotherms. This is due to adaptations such as low resting metabolism, muscle contractions, and sunning.

Reptiles have lungs and must breathe air, even aquatic and semi-aquatic species.

A three-chambered heart pumps blood throughout the system.

When the dinosaurs ruled the Earth millions of years ago, mammals evolved from a niche of reptiles deemed synapsids.

These organisms are known as *para mammals*, *proto-mammals*, or *mammal-like reptiles*.

Mammals

Mammals (class *Mammalia*) diverged from true reptiles (sauropsids) and became true mammals, gradually shedding their reptilian features for hair, internal amniotic eggs, and mammalian characteristics.

Although dinosaurs dominated for some time, mammals became competitors for land resources and predated some smaller dinosaurs.

Mammals are diverse organisms but share common apomorphies, including:

> *hair*, *varied teeth* (or *heterodonty*),
>
> *warm-blooded metabolism* (or *endothermy*), and
>
> *mammary glands* to feed young with milk.

Nearly all mammals give live birth, except for the monotremes, which lay eggs like their reptilian ancestors.

Most mammals are *placental*, with offspring developing in the *uterus,* nourished by a *placenta.*

Marsupials carry their offspring in an external womb, the pouch.

Birds

Birds (class *Aves*) descended from reptiles as a highly diverse class with thousands of species.

Class Aves divides into:

> *paleognaths* (*flightless* and *weak-flying birds*) and

> *neognaths* (all other birds).

Birds have many unique apomorphies which separate them from their reptilian kin.

Their reptilian ancestry is evident in scales, feathers (scale derivatives), and amniotic eggs.

Unlike reptiles, most birds have *bills* rather than teeth.

They are adapted for *flight*, having lightweight bones, forelimbs modified into wings, large flight muscles, and feathers.

To meet the high energy demands of flight, birds have a high metabolism and sophisticated respiratory and circulatory systems.

Endothermy

A large amount of heat generated makes birds endotherms like mammals, an analogous trait.

Endothermy was not found in common ancestors but evolved separately in birds and mammals.

A similar example is the ability of flight found in birds and bats.

Birds have more powerful lungs and massive four-chambered hearts than mammals, allowing them to circulate oxygen efficiently.

Respiration has the additional benefit of filling the spaces within their bones, contributing to their buoyancy.

Many birds are such strong fliers that they can migrate thousands of miles yearly.

Birds may be insectivorous, herbivorous, carnivorous, omnivorous, or eat carrion.

The wide variation in bird morphology is a testament to their diversification into many niches.

Notes for active learning

Ontogeny and Phylogeny Relationships

Recapitulation theory

Ontogeny is the development of an organism through life, from a zygote to the adult form.

Phylogeny is the development of a species over evolutionary time.

Historically, evolutionary biologists subscribed to the *"ontogeny recapitulates phylogeny."*

Recapitulation theory proposes that the development observed (i.e., *ontogeny*) reflects evolution (i.e., *phylogeny*).

For example, the notochord classifies vertebrates. A vertebrate embryo has a notochord in early development, but vertebrae later replace it.

Similarly, the chordate ancestor had a notochord replaced by vertebrae during the evolution of vertebrates. In this case, the theory holds.

However, the notion that ontogeny recapitulates phylogeny is a refuted exaggeration.

If true, a human would appear as a mature invertebrate (e.g., fish, reptile) and then an ancestral mammal until it developed into a mature human. This does not occur.

Instead, the human embryo diverges from generalized forms as it matures.

It does not resemble a fish during development, but merely that an early embryo cannot be visually identified as either fish or human.

Embryology

Embryology is separate from the *recapitulation theory* to provide valuable insights into developing embryos.

It cannot be concluded that a human embryo evolved from a bird ancestor merely because a human embryo and a bird embryo look similar until several weeks after fertilization.

Since they look similar, it can be inferred that bird embryos are more related to humans than fish.

Ontogeny reveals evolutionary relatedness but not the evolution order or derived characteristics.

Classification and diversity summary

Evolution	How modern organisms have descended from ancient organisms
Common ancestor	An ancestor shared by two or more descendant species
Fossil	Preserved remains of ancient organisms
Homologous structure	Structure similar in different species due to common ancestry
Vestigial structure	Structure is non-functional or reduced in function
Analogous structure	Structure evolved independently in organisms from living in similar environments or experiencing similar selective pressures
Embryology	Study of embryos and their development
Biogeography	Studies of where organisms live and where ancestors lived

Phylum summary

Phylum	Common names	Germ layers	Body symmetry	Gut openings	Coelom	Embryonic development
Porifera	sponges	-	none	0*	-	-
Cnidaria	jellyfish, corals	2	radial	1	-	-
Platyhelminthes	flatworms	3	bilateral	1	acoelomate	-
Rotifera	rotifers	3	bilateral	2	pseudo-coelomate	-
Nematoda	roundworm	3	bilateral	2	pseudo-coelomate	-
Annelida	segmented worms	3	bilateral	2	coelomate	protosome
Mollusca	clams, snails, octopuses	3	bilateral	2	coelomate	protosome
Arthropoda	crustaceans, spiders, insects	3	bilateral	2	coelomate	protosome
Echinodermata	sea stars, sea urchins	3	radial	2	coelomate	deuterostome
Chordata	vertebrates	3	bilateral	2	coelomate	deuterostome

* *Amoebocytes carry out digestion*

- Characteristic does not apply to this phylum.

Notes for active learning

Notes for active learning

Notes for active learning

PRACTICE QUESTIONS
&
DETAILED EXPLANATIONS

Page intentionally left blank

Practice Questions: Evolution & Natural Selection

1. During his trip to the Galapagos Islands in 1835, Darwin discovered that:

   A. local fossils were not related to contemporary organisms of that region

   B. all organisms were related

   C. organisms in tropical regions were closely related and independent of the region

   D. several species of finches varied from island to island

   E. organisms had no relationship with others

2. In evolutionary terms, organisms belonging to which category would be the most similar?

   A. genus

   B. family

   C. order

   D. kingdom

   E. class

3. What is the source of genetic variation for natural selection?

   A. gene duplication

   B. mitosis

   C. meiosis

   D. heterozygosity

   E. mutation

4. In modern evolutionary theory, chloroplasts probably descended from:

   A. free-living cyanobacteria

   B. red algae

   C. mitochondria

   D. aerobic prokaryote

   E. eukaryotic cells

5. Speciation is the evolution of new, genetically distinct populations from a common ancestral stock by:

   I. random mutation

   II. geographic isolation

   III. reduction of gene flow

   A. I and III only

   B. II and III only

   C. I and II only

   D. I, II, and III

   E. III only

6. Darwin (1809-1882) hypothesized that the mechanism of evolution involves:

A. selective pressure

B. epigenetics

C. natural selection

D. selective breeding

E. random selection

7. The complexity of the organisms that exist on earth today is the result of:

    I. multicellularity

    II. photosynthesis

    III. eukaryotic cell development

A. I only

B. II only

C. I and II only

D. II and III only

E. I, II and III

8. The Archaean Eon contains the oldest known fossil record dating to about:

A. 1.5 billion years

B. 2.0 million years

C. 7,000 years

D. 3.5 billion years

E. 7 billion years

9. Urey-Miller experiment in 1952 demonstrates that:

A. life may have evolved from inorganic precursors

B. humans have evolved from photosynthetic cyanobacteria

C. life existed on prehistoric earth

D. small biological molecules cannot be synthesized from inorganic material

E. the early earth lacked oxygen

10. To ensure the survival of their species, animals that do not care for their young:

A. have protective coloring

B. produce many offspring

C. lay eggs

D. can live in water and on land

E. have internal fertilization

11. Which best describes the relationship between nitrogen-fixing bacteria that derive their nutrition from plants and the plants that benefit from the nitrogen supplied by the bacteria?

A. parasitism

B. commensalism

C. mutualism

D. mimicry

E. amensalism

12. Which selection type is likely to lead to speciation?

A. sexual selection

B. stabilizing selection

C. directional selection

D. disruptive selection

E. nondirectional selection

13. Similarity among the embryos of fish, amphibians, reptiles, and humans is evidence of:

A. genetic drift

B. sexual selection

C. analogous traits

D. common ancestry

E. genetic equilibrium

14. Selective breeding of soybeans by humans has genetically altered the soybean so that it could not survive in the wild without human intervention. The soybean population is controlled; most soybean seeds are eaten or spoiled. The relationship between humans and soybeans is best described as follows:

A. commensalism, because both species benefit

B. parasitism, because humans benefit, and soybeans are harmed

C. commensalism, because there is no benefit to either species

D. commensalism, because humans benefit, and soybeans are neither benefited nor harmed

E. mutualism, because both species benefit

15. Natural selection leads to:

A. larger population

B. population most adapted to its present environment

C. phenotypic diversity

D. broad genetic variation

E. population well adapted to changes in the environment

16. From an evolutionary perspective, which cell property is paramount?

A. passing genetic information to progeny

B. containing a nucleus

C. containing mitochondria

D. interacting with other cells

E. extracting energy from the environment

17. Inherited traits determined by elements of heredity are:

A. crossing over

B. homologous chromosomes

C. locus

D. p elements

E. genes

18. Which statement is correct about the evolutionary process?

A. Organisms develop traits that they need for survival

B. Mutations decrease the rate of evolution

C. Natural selection works on traits that cannot be inherited

D. Natural selection works on existing genetic variation within a population

E. Organisms evolve due to natural selection

19. Genetic drift results from:

A. environmental change

B. genetic diversity

C. probability

D. sexual selection

E. genetic mutation

20. The likely result of polygenic inheritance is:

A. human height

B. blood type

C. freckles

D. polydactyly (extra digits)

E. color of the iris

21. Which is the earliest form of life to evolve on Earth?

A. plants

B. eukaryotes

C. prokaryotes

D. protists

E. fish

22. If two animals produce viable, fertile offspring under natural conditions, it can be concluded that:

A. for any given allele, they both have the same gene

B. their blood types are compatible

C. they both have haploid somatic cells

D. they both are from the same species

E. none of the above

23. The population's total collection of alleles at any one time makes up its:

A. phenotype

B. biodiversity

C. gene pool

D. genotype

E. genetic mutations

24. A pivotal point in Darwin's explanation of evolution is that:

A. biological structures inherited are better suited to the environment through constant use

B. mutations that occur are those that help future generations fit into their environments

C. mutations develop based on the use/disuse of physical traits

D. genes change to help organisms cope with problems encountered within their environments

E. any trait that confers even a slight increase in the probability that its possessor will survive and reproduce will be strongly favored and will spread in the population

25. If the allele frequency in a population is 0.7, what is the alternate allele frequency?

A. 0.14

B. 0.21

C. 0.30

D. 0.4

E. 0.42

26. Which statement about evolution is CORRECT?

A. Darwin's theory of natural selection relies solely on environmental conditions

B. Darwin's theory explains the evolution of man from present-day apes

C. Darwin's theory of natural selection relies solely on genetic mutation

D. Lamarck's theory of use and disuse adequately describes why giraffes have long necks

E. Natural selection is when random mutations are selected for survival by the environment

27. A significant genetic drift may be prevented by:

A. random mutations

B. large population size

C. small population size

D. genetic variation

E. lack of migration

28. Which statement would apply to a population that survived a bottleneck and recovered to its original size?

    A. It is subject to genetic drift
    B. It is less likely to become extinct than before the bottleneck
    C. It has more genetic variation than before the bottleneck
    D. It has less genetic variation than before the bottleneck
    E. None of the above applies

29. Difference between the founder effect and a population bottleneck is that founder effect:

    A. only occurs in island populations
    B. requires the isolation of a small group from a larger population
    C. requires a large gene pool
    D. involves sexual selection
    E. always follows genetic drift

30. Which compound was likely unnecessary for the origin of life on Earth?

    A. carbon
    B. O_2
    C. hydrogen
    D. H_2O
    E. nitrogen

31. In a population of humans, there is a higher rate of polydactyly (extra fingers or toes) than in the human population. The likely explanation is:

    A. bottleneck effect
    B. founder effect
    C. sexual selection
    D. natural selection
    E. missense mutation

32. Gene flow occurs through:

    A. random mutations
    B. bottleneck effect
    C. founder effect
    D. directional selection
    E. migration

33. The first living organisms on Earth derived their energy from:

A. eating dead organisms

B. eating organic molecules

C. the sun

D. eating each other

E. all the above

34. Evolutionary fitness measures:

A. population size

B. reproductive success

C. gene pool size

D. genetic load

E. migration rate

35. Which scientist would explain the hawk's lost flying ability with a theory that "since the hawk stopped using its wings, the wings became smaller, and this acquired trait was passed on to the offspring"?

A. Lamarck

B. de Vries

C. Mendel

D. Darwin

E. Morgan

36. Which example is a likely result of directional selection?

A. Increased number of cat breeds

B. Python snakes with assorted color patterns exhibit different behavior when threatened

C. Female *Drosophila* select usual yellowish-grey colored males over less common, yellow-colored males

D. Non-poisonous butterflies evolving color changes to look like poisonous butterflies

E. None of the above

37. The genetic variation would be decreased by:

A. genetic drift

B. balancing selection

C. directional selection

D. sexual selection

E. stabilizing selection

38. Asexually reproducing species have a selective disadvantage over sexually reproducing species because sexual reproduction:

    A. always decreases an offspring's survival ability

    B. decreases the likelihood of mutations

    C. creates novel genetic recombination

    D. is more energy efficient

    E. always increases an offspring's survival ability

39. Which example is a likely result of stabilizing selection?

    A. Female *Drosophila* select yellowish-grey colored male mates over less common, yellow-colored males

    B. Increased number of different breeds within one species

    C. Non-poisonous butterflies evolving color changes to look like poisonous butterflies

    D. Female birds that lay an intermediate number of eggs have the highest reproductive success

    E. none of the above

40. According to one theory about the origins of life, these molecules were formed by purines, pyrimidines, sugars, and phosphates combining:

    A. lipids

    B. carbohydrates

    C. nucleosides

    D. proteins

    E. nucleotides

41. Which example is a likely result of sexual selection?

    A. Female deer choosing to mate with males that have the biggest antlers

    B. Python snakes with distinct color patterns exhibit different behavior when threatened

    C. Non-poisonous butterflies evolving color changes to look like poisonous butterflies

    D. Increased number of different breeds within one species

    E. Lions experience a founder effect

42. Some fossils from before the Cambrian Explosion are embryos, suggesting that the organisms:

    A. reproduce sexually

    B. exhibit cephalization

    C. reproduce asexually

    D. have bilateral symmetry

    E. are protostomes

43. Cambrian Explosion resulted in the evolution of the first:

A. bacteria

B. land animals

C. dinosaurs and mammals

D. representatives of most animal phyla

E. plants

44. A primate's ability to hold objects in its hands or feet is an evolutionary development necessary for it to:

A. create elaborate social systems

B. consume food

C. use simple tools

D. walk upright

E. escape predation

45. Suppose a paleontologist discovers a fossil skull that he believes might be distantly related to primates. Unlike true primates, the face is not flat, and the eyes do not face entirely forward. The paleontologist would conclude that the animal lacked the ability to:

A. manipulate tools

B. judge the location of tree branches

C. form extended family groups

D. grip branches precisely

E. harvest crops

46. When an animal's environment changes, sexual reproduction improves a species' ability to:

A. increase its numbers rapidly

B. produce genetically identical offspring

C. react to new stimuli

D. adapt to new living conditions

E. reduce genetic diversity

47. Researchers have concluded that Dikika Baby was a better climber than modern humans. This evidence suggests that Dikika Baby's relatives may have spent part of their time:

A. hunting in groups

B. waging war on other troops

C. using tools

D. in trees

E. in the Rocky Mountains

48. Fossil evidence indicates that *Australopithecus afarensis:*

A. was bipedal

B. appeared later than *Homo ergaster*

C. was primarily a meat-eater

D. had a large brain

E. was a hunter and gatherer

49. One way in which reptiles are adapted to fully terrestrial life is:

A. endothermy

B. viviparous development

C. external fertilization

D. placenta

E. amniotic egg

50. Researchers concluded from the leg bones of the fossil known as Lucy that it was bipedal. Which of the following indicates that this hominine was bipedal?

A. bowl-shaped pelvis

B. opposable thumbs

C. skull with a flat face

D. broad rib cage

E. flexible spinal column

51. Which statement is true of *Homo sapiens?*

A. They replaced *Homo habilis* in the Middle East

B. They became extinct about 1 million years ago

C. They have been Earth's only hominine for the last 24,000 years

D. They evolved after the Cro-Magnons

E. They replaced *Homo habilis* in Europe

52. Which was a unique characteristic of Neanderthals?

A. Producing tools from bones and antlers

B. Burying their dead with simple rituals

C. Making sophisticated stone blades

D. Producing cave paintings

E. Members joined into groups

Notes or active learning

Notes or active learning

Detailed Explanations: Evolution & Natural Selection

Answer Key

1: D	11: C	21: C	31: B	41: A	51: C
2: A	12: D	22: D	32: E	42: A	52: B
3: E	13: D	23: C	33: C	43: D	
4: A	14: E	24: E	34: B	44: C	
5: D	15: B	25: C	35: A	45: B	
6: C	16: A	26: E	36: D	46: D	
7: E	17: E	27: B	37: E	47: D	
8: D	18: D	28: D	38: C	48: A	
9: A	19: C	29: B	39: D	49: E	
10: B	20: A	30: B	40: E	50: A	

1. D is correct.

Charles Darwin (1809-1882), during his trip to the Galapagos Islands in the 1830s, observed several species of finches that differed among islands.

2. A is correct.

Evolutionary relationships classify organisms:

$$\text{Kingdom} \rightarrow \text{Phylum} \rightarrow \text{Class} \rightarrow \text{Order}$$

$$\rightarrow \text{Family} \rightarrow \text{Genus} \rightarrow \text{Species}$$

Largest group (i.e., kingdom) divides into smaller subdivisions.

Each smaller group has common characteristics. Of the answers, a genus is the smallest subdivision and organisms in the same genus are more similar than organisms classified as the same family, order, class, or kingdom.

3. E is correct.

Mutations are the source of genetic variation for natural selection.

4. A is correct.

Chloroplast is a plant cell organelle that may have originated from cyanobacteria through *endosymbiosis*.

Endosymbiotic theory proposes that a eukaryotic cell engulfed a photosynthesizing cyanobacterium about one billion years ago.

Photosynthesizing cyanobacterium became a permanent resident in the cell because it escaped the phagocytic vacuole in which it was contained.

Chloroplasts contain their DNA and ribosomes (like prokaryotic ribosomes) and undergo autosomal replication (i.e., replicating independently of the cell cycle), supporting this theory.

Cyanobacteria obtain energy by photosynthesis and have the color of the bacteria (i.e., blue).

Named blue-green algae, this is a misnomer as cyanobacteria are prokaryotes, while algae are eukaryotes.

By producing oxygen as a by-product gas of photosynthesis, cyanobacteria converted Earth's early reducing atmosphere into an oxidizing one, dramatically changing the composition of life by stimulating biodiversity and leading to the near-extinction of oxygen-intolerant organisms.

Endosymbiotic theory proposes that chloroplasts in plants, mitochondria in eukaryotes, and eukaryotic algae evolved from prokaryotic cyanobacterial ancestors.

Mitochondria evolved from free-living prokaryotic heterotrophs that, similarly to chloroplasts, entered eukaryotic cells and established a symbiotic relationship with the host (the theory of *endosymbiosis*).

Mitochondria contain circular DNA (similar in size and composition to prokaryotes), ribosomes (like prokaryotic ribosomes), and unique proteins in the organelle membrane (similar in composition to prokaryotes) and replicate independently, providing further evidence to support this theory.

5. D is correct.

Evolution is the long-term changes in a population's gene pool caused by environmental selection pressures.

I: *random mutation* creates new alleles selected for (or against) as a phenotypic variation for natural selection (i.e., survival of the fittest).

II: *reproductive isolation* of a population is a crucial component of speciation.

Geographically isolated populations often diverge to yield reproductive isolation leading to speciation.

III: *speciation* might happen in a population with no specific extrinsic barrier to gene flow.

For example, a population extends over a broad geography, and mating in the population is not random.

Individuals in the far west would have zero probability of mating with individuals in the far east range.

This results in reduced *gene flow* but not total isolation.

Such a situation may (or not) be sufficient for speciation.

Speciation would be promoted by different selective pressures at opposite ends of the range, which would alter gene frequencies in groups at different ranges so they would not be able to mate if they were reunited.

6. C is correct.

Darwin (1809-1882) hypothesized that the mechanism of evolution involves natural selection.

7. E is correct.

Prokaryotes are primarily *unicellular* organisms, although a few, such as mycobacterium, have multicellular stages in their life cycles or create large colonies like cyanobacteria.

Eukaryotes are often *multicellular* and are typically much *larger than prokaryotes*.

Eukaryotes have *internal membranous structures* (i.e., organelles) and a cytoskeleton composed of microtubules, microfilaments, and intermediate filaments, which are essential for cellular organization and shape.

8. D is correct.

Archaean Eon contains the oldest known fossil record, dating to about 3.5 billion years.

9. A is correct.

Urey and Miller demonstrated in 1953 that organic molecules might be created from inorganic molecules under primordial earth conditions.

Urey-Miller designed an experiment that simulated conditions thought to be on the early Earth and assessed for the occurrence of chemical origins of life.

Urey-Miller did not prove the existence of life on earth.

However, experiments showed over 20 amino acids produced in Miller's experiments.

10. B is correct.

Nature uses many methods for fertilization, development, and care of offspring.

Internal fertilization involves internal development and much care for offspring.

These organisms (e.g., humans and many mammals) produce few offspring, but a substantial percentage reach adulthood.

External fertilization involves external development and little care for offspring.

These organisms (e.g., many species of fish) produce large numbers of sperm and eggs because few sperm and eggs interact to produce a zygote.

Few zygotes survive without physical protection from predators, and millions of eggs and sperm must be released to perpetuate the species.

A: *protective coloring* might help survival against predators, but it is not related to caring for the young.

C: *laying eggs* is not related to not caring for the young (most birds care for their young).

D: not relevant to the survival rate of the offspring.

11. C is correct.

Mutualism is when two species live in close association and both benefit.

Parasitism is a relationship between two organisms where one organism benefits while the other is harmed.

Commensalism involves a relationship between organisms where one benefits without affecting another.

Mimicry (in evolutionary biology) is the similarity of species, which protects one or both and occurs when a group (the mimic) evolves from sharing common characteristics with another group (the models). Similarity can be in appearance, behavior, sound, scent, or location, mimicking similar places to their models.

Amensalism is a relationship where one species is inhibited or wholly obliterated while the other is unaffected.

For example, consider a sapling growing under the shadow of a mature tree. The mature tree can deprive the sapling of sunlight, rainwater, and soil nutrients. Throughout the process, the mature tree is unaffected.

Additionally, if the sapling dies, the mature tree will gain nutrients from the decaying sapling.

Since the nutrients become available due to the dead sapling's decomposition (as opposed to the living sapling), this would not be a case of parasitism.

12. D is correct.

Disruptive selection shifts allele frequencies towards variants of extremes, leading to two subpopulations.

These subpopulations do not favor the intermediate variants of the original population and grow disparate over time, likely leading to speciation.

A: *sexual selection* gives an individual an advantage in finding a mate.

Sexual selection often acts opposite to the effects of natural selection (e.g., a brighter plume may make a bird vulnerable to predators but gives it an advantage during reproduction).

It is hypothesized that sexual selection leads to sexual dimorphism (i.e., phenotypic differences between males and females of the same species).

B: *stabilizing selection* is a shift in phenotypes towards intermediates by disfavoring variants at extremes.

Stabilizing selection reduces variation within the species and perpetuates similarities between generations.

C: *directional selection* is the population shift in allele frequencies towards variants of one extreme.

13. D is correct.

Similarity among the embryos of fish, amphibians, reptiles, and humans is evidence of *common ancestry*.

14. E is correct.

Mutualism (i.e., symbiotic) is a relationship between two organisms where both benefit.

For example, soybeans depend on humans to survive while humans are provided with food.

Both species benefit from a mutualism relationship.

Commensalism is a class of relationships where one organism benefits without affecting the other.

Mutualism (i.e., symbiotic) is a relationship between two organisms where both benefit.

Parasitism is a relationship between two organisms where one benefits while the other is harmed.

15. B is correct.

Natural selection leads to a population most adapted (*fittest*) to its current environment.

16. A is correct.

From an evolutionary perspective, *passing genetic information* to progeny is paramount.

17. E is correct.

Inherited traits determined by heredity are genes.

18. D is correct.

The evolutionary process of natural selection *works on existing genetic variation* within a population.

19. C is correct.

Genetic drift changes allele frequency within a population due to random sampling.

Allele frequency of a population is the fraction of the same alleles.

Offspring have the same alleles as parents, and probability determines if an individual survives and reproduces.

Genetic drift may cause allele variants to disappear and thereby reduce genetic variation.

20. A is correct.

Human height characteristics result from *polygenic* (i.e., many genes contribute) inheritance.

21. C is correct.

The oldest known *fossilized prokaryotes* were laid down approximately 3.6 billion years ago, about 1 billion years after the formation of the Earth's crust.

Eukaryotes appear in the fossil record later and may have formed from the aggregation of multiple prokaryotes.

The oldest known *fossilized eukaryotes* are about 1.7 billion years old.

However, some genetic evidence suggests eukaryotes appeared as early as 3 billion years ago.

Protists are a large and diverse group of eukaryotic microorganisms that belong to the kingdom Protista.

There have been attempts to remove the Protista kingdom from the taxonomy, but it is commonly used.

Some professional organizations and institutions prefer the name Protista.

Protists are unicellular, or they are multicellular without specialized tissues.

continued...

Besides their relatively simple levels of organization, protists do not have much in common.

The straightforward cellular organization distinguishes protists from other eukaryotes (e.g., fungi and animals).

Protists live in almost any environment with liquid water.

Many protists, such as algae, are photosynthetic and are vital primary producers in ecosystems, particularly in the ocean as part of the plankton.

Protists include pathogenic species, such as the kinetoplastid *Trypanosoma brucei,* causing sleeping sickness, and species of the apicomplexan *Plasmodium,* causing malaria.

22. D is correct.

Morphological or physical similarity between organisms is insufficient for classifying the same species.

Species are organisms that mate and produce *viable* and *fertile* offspring (give rise to additional offspring).

Organisms appearing different (e.g., dog breeds) are the same species if they mate and produce *fertile offspring*.

A: organisms can be the same species yet have different gene varieties (i.e., alleles) at the same gene locus.

For example, blue-eyed and green-eyed individuals can mate and produce fertile offspring.

B: *no relationship* between producing viable, fertile offspring and blood type.

C: *somatic cells* are body cells, not reproductive cells (i.e., gametes) or haploid (i.e., 1N or monoploid).

Somatic cells are diploid (2N) and undergo *mitosis* for cell division.

Gametes are haploid (1N) and undergo *meiosis* for cell division.

23. C is correct.

Gene pool is the population's total collection of alleles.

24. E is correct.

A pivotal point in Darwin's explanation of evolution is that any trait that confers an increase in the *probability* of its possessor surviving and reproducing is favored and spread through the population.

25. C is correct.

If an allele's frequency in a population is 0.7, the alternate allele is 0.3.

26. E is correct.

Natural selection is how mutations are selected for (or against) in the environment.

If the resulting phenotype offers some degree of fitness, the genes are passed to the next generation.

A: *natural selection* includes selection pressures and needs a population with genetic variation (random mutations) to select the fittest organisms.

C: *Darwin's theory* depends on more than mere mutations. It is based on over-reproduction.

Offspring are selected for (or against) when their genetic makeup fits the local environment.

Organisms most fit (for their environment) and pass their genes (i.e., gametes) to the next generation.

The result is survival (greatest reproduction) of the *fittest organisms in the environment.*

D: *Lamarck* proposed that if traits were used (e.g., stretching of a giraffe's neck), these acquired traits are passed to the next generation.

Acquired characteristics (phenotypic changes) do not affect the genes and are not passed to the next generation via the gametes (e.g., sperm and egg).

27. B is correct.

Fewer copies of an allele magnify the effect of *genetic drift.*

Genetic drift is negligible when there are many copies of an allele.

28. D is correct.

A population that survived a bottleneck and recovered to its original size has *less genetic variation* than before.

29. B is correct.

Founder effect and *population bottleneck* differ because the founder effect requires isolating a small group from a larger population.

30. B is correct.

Life originated in an atmosphere with *little or no oxygen.*

Until the Great Oxygenation Event (GOE) about 2.4 billion years ago, there was no atmospheric free oxygen.

Cyanobacteria appeared about 200 million years before the GOE began producing oxygen by photosynthesis.

Before the GOE, dissolved iron or organic matter chemically captured free oxygen.

The GOE was when oxygen sinks became saturated with oxygen produced by cyanobacterial photosynthesis.

Excess free oxygen accumulated in the atmosphere after the GOE.

Free oxygen is toxic to obligate anaerobic organisms. The rising concentrations eradicated most of Earth's anaerobic organisms.

Cyanobacteria were responsible for one of Earth's most significant extinction events.

Free oxygen reacts with atmospheric methane (a greenhouse gas), which is oxidized to CO_2 and H_2O.

Free oxygen has been an essential constituent of the atmosphere ever since.

Periods with much oxygen in the atmosphere are associated with the rapid development of animals.

Today's atmosphere contains about 21% oxygen, which is enough for the rapid development of animals.

31. B is correct.

Founder effect is the likely explanation that a population of humans has a higher rate of polydactyly (extra fingers or toes) than the human population.

32. E is correct.

Gene flow occurs through migration.

33. C is correct.

The first organisms had nothing else to eat. Even with millions of organisms, one organism must eat many, which would have exhausted the supply.

34. B is correct.

Evolutionary fitness measures *reproductive success* (i.e., hereditary succession of gene pools).

35. A is correct.

Lamarck's theory of evolution proposed that enhanced structures (e.g., a giraffe's neck) arise based on use.

Lamarck (1744-1829) had a false premise that favorable characteristics obtained by use were inheritable.

For example, giraffes stretch their necks to reach leaves on branches.

Lamarck proposed that offspring inherit the trait (longer necks) based on use. However, only changes in the DNA of gametes (e.g., egg or sperm) are inherited.

B: de Vries (1848-1935) confirmed Mendel's observations with different plant species.

C: Mendel (1822-1884) defined classical genetics by experiments with inheritable traits (e.g., seed color, wrinkled vs. round seeds) in pea plants.

Mendel described the principles of dominance, segregation, and independent assortment. The unit of inheritance (i.e., genes) was unknown.

D: *Darwin's Theory of Natural Selection* states that *environmental pressures select the fittest organism* to survive and reproduce.

Natural Selection by Mendel is based on six principles.

1) **Overpopulation**: more offspring are produced than can survive with insufficient food, air, light, and space to support the population;

2) **Variations**: offspring have different characteristics (i.e., variations) than the population.

Darwin did not know why, but de Vries later suggested that *genetic mutations* cause variations.

Some mutations are *beneficial* fittest), but most are *detrimental*;

3) **Competition**: developing populations compete for necessities (e.g., food, air, light).

Many young die, while the number of adults remains constant for generations;

4) **Natural selection**: some organisms have phenotypes (i.e., variations) conferring an advantage over others;

continued...

5) **Inheritance of variations**: individuals reproduce and transmit favorable phenotypes.

Favored alleles (i.e., variations of genes) eventually dominate the gene pool;

6) **Evolution of new species**: natural selection (fittest) propagates favorable genes and phenotypes over generations.

Favorable changes result in significant changes in the gene pool for the evolution of a new species (i.e., organisms that reproduce and yield fertile offspring).

E: Morgan (1866-1945) induced mutation in *Drosophila,* studied the inheritance of these mutations, and described sex-linked inheritance (i.e., Morgan unit for the frequency of recombination within an organism).

36. D is correct.

Non-poisonous butterflies evolve color changes to look like poisonous butterflies in *directional selection.*

37. E is correct.

Genetic variation would likely be decreased by stabilizing selection.

38. C is correct.

Asexual reproduction is more efficient than sexual reproduction in the number of offspring produced per reproduction, in the amount of energy invested in this process, and in the amount of time invested in the development of the young (before and after birth).

Asexual reproduction relies on *genetic mutation* for phenotypic variability to be passed to future generations since it produces genetic clones of the parent.

Sexual reproduction involves the process of meiosis – two rounds of cell division and the likelihood of cross-over (during prophase I).

A: new phenotype may be disadvantageous *or* advantageous.

B: a much greater probability of a mutation occurring with sexual reproduction (i.e., at the chromosomal level), known as chromosomal aberrations.

D: *sexual reproduction* requires more *energy* and *time* per progeny (i.e., offspring).

Species reproducing sexually have a selective advantage because of *gene recombination* during fertilization.

continued...

Fusing two genetically unique nuclei (haploid sperm and haploid egg nucleus) yields a unique (2N) zygote.

Fertilizing two unique gametes introduces phenotypic variability into a population.

This genetic and phenotypic variability may benefit or harm individuals in their environment.

Advantageous phenotypes survive and pass genes to future generations consistent with natural selection (i.e., survival of the fittest).

39. D is correct.

Birds laying an intermediate number of eggs have the highest reproductive success, likely resulting from stabilizing selection.

40. E is correct.

Nucleotides are sugar (ribose for RNA or deoxyribose for DNA), a phosphate group, and a base.

Nitrogenous bases are guanine, adenine, cytosine, thymine (for DNA), or uracil (for RNA).

Adenine and guanine are purines, while thymine, cytosine, and uracil are pyrimidines.

A: *lipids* (i.e., fats) are composed of glycerol (i.e., 3-carbon chain) and (up to) three fatty acids.

B: *monosaccharides* are the monomers (e.g., glucose, fructose) of carbohydrates (e.g., glycogen, cellulose).

C: *nucleosides* have a nitrogenous base and sugar (without phosphate).

Nucleotides have a phosphate group, nitrogenous base, and sugar.

D: amino acids are monomers for proteins that do not contain sugar, phosphate or bases.

41. A is correct.

Sexual selection likely results from a female deer choosing to mate with males with the biggest antlers.

42. A is correct.

Embryos indicate that fossilized organisms reproduce sexually.

The development of an embryo involves two distinct reproductive cells (i.e., gametes) fusing, characteristic of sexual reproduction.

This single-cell zygote results from fertilizing the female egg cell with a male sperm cell.

43. D is correct.

Cambrian Explosion occurred 542 million years ago with the rapid appearance of most major animal phyla. The fossil record shows that this was accompanied by significant diversification of other organisms.

Before about 580 million years ago, most organisms were simple, composed of individual cells occasionally organized into colonies.

Over the following 70-80 million years, the rate of evolution accelerated by order of magnitude, and the diversity of life became similar to today.

44. C is correct.

During evolution, primates needed to walk on two legs (i.e., *bipedal*) to free their hands for other tasks.

Hands were used in creating weapons and tools to gather food and protect the young.

45. B is correct.

Eyes facing forward allow two fields of vision to overlap, enabling the perception of depth.

Animals without facing forward eyes could not likely judge the distance to objects (e.g., tree branches) due to flawed depth perception.

As primates moved into trees to escape predators, they needed to navigate tree branches, meaning evolution favored good depth perception.

46. D is correct.

Sexual reproduction increases fitness and allows for modifications in traits, which can help adapt to new living conditions in a changing environment.

Sexual reproduction adds to the diversity of the offspring.

All other answer choices characterize asexual reproduction.

47. D is correct.

Arboreal locomotion is the movement of animals in *trees*.

Some animals have evolved to move in habitats where trees are present.

Animals may only scale trees occasionally or become exclusively arboreal.

48. A is correct.

Australopithecus afarensis is an extinct species with hip, knee, and foot morphology distinctive to bipedalism.

Footprints show *Australopithecus afarensis* walked with an upright posture and a decisive heel strike, which is early evidence of bipedalism.

49. E is correct.

Amniotic eggs are laid on land by certain reptiles, birds, and mammals.

Amniotic eggs differ from *anamniotic* eggs, which are typically laid in water (by fish and amphibians).

50. A is correct.

Gluteus muscles of the pelvis are important for propulsion and stability while walking.

Bipedal humans are different from quadrupedal apes in the lateral orientation of the ilium, which constitutes the bowl-shaped pelvis.

51. C is correct.

Homo sapiens evolved after the *Homo erectus* and are *modern humans*.

52. B is correct.

Neanderthals are closely related to modern humans, with differences based on DNA by 0.3%, but twice the greatest DNA difference among contemporary humans.

Genetic evidence suggests Neanderthals contributed to the DNA of anatomically modern humans, probably through interbreeding between 80,000 and 30,000 years ago.

Recent evidence suggests that Neanderthals practiced burial behavior and buried their dead.

Archaeological remains by Neanderthals include bones and stone tools.

Notes for active learning

Notes for active learning

Practice Questions: Classification & Diversity

1. *Canis lupus* is a Canidae family wolf. Which statement is accurate about its members?

   A. They may be classified as *lupus* but not *Canis*

   B. They may be classified as *Canis* but not Canidae

   C. More are classified as *lupus* than *Canis*

   D. More are classified as Canidae than *Canis*

   E. None of the above

2. Two species of the same order must be members of the same:

   A. class

   B. habitat

   C. genus

   D. family

   E. subfamily

3. Homology serves as the evidence of:

   A. balancing selection

   B. biodiversity

   C. genetic mutation

   D. sexual selection

   E. common ancestry

4. Which statement is correct about orthologous genes?

   A. They are related through speciation

   C. They are repetitive

   B. They are house-keeping genes

   D. They are related via gene duplication within species

   E. They are viral oncogenes

5. Which given pair represents homologous structures?

   A. Mouth of a fly and beak of a hummingbird

   B. Wings of a dragonfly and a blue jay

   C. Forelimb of a human and of a dog

   D. Wings of a pigeon and a bat

   E. Forelimb of a raccoon and the hind limb of a wolf

6. Humans belong to the order:

A. hominidae

B. primata

C. chordata

D. vertebrata

E. sapiens

7. All are members of the phylum Chordata EXCEPT:

A. birds

B. ants

C. tunicates

D. apes

E. snakes

8. All statements regarding the phylum Echinodermata are true, EXCEPT:

A. echinoderms are heterotrophs

B. echinoderms are invertebrates

C. echinoderms reproduce sexually

D. phylum includes starfish and sea urchins

E. phylum includes crayfish

9. Two species of the same phylum must be members of the same:

A. order

B. kingdom

C. genus

D. class

E. family

10. *Homo sapiens* belong to the phylum:

A. Vertebrata

B. Homo

C. Mammalia

D. Chordata

E. sapiens

11. Which organism is a chordate but not a vertebrate?

A. lizard

B. lancelet

C. shark

D. lamprey eel

E. none of the above

12. Which represents the levels of complexity at which life is studied, from the simplest to the most complex?

A. cell, tissue, organ, organism, population, community

B. community, population, organism, organ, tissue, cell

C. cell, organ, tissue, organism, population, community

D. cell, tissue, organ, population, organism, community

E. tissue, organ, cell, population, organism, community

13. Which statement is TRUE for the notochord?

A. It is present in chordates during embryological development

B. It is always a vestigial organ in chordates

C. It is present in all adult chordates

D. It is present in all echinoderms

E. It is part of the nervous system of all vertebrates

14. All are present in members of the phylum Chordata at some point in their life cycle, EXCEPT:

A. pharyngeal slits

B. tail

C. notochord

D. dorsal neural tube

E. backbone

15. Which are likely two organisms of the same species?

A. Two South American iguanas which mate in different seasons

B. Two migratory birds nesting on the different Indonesian Islands

C. Cabbage in Alabama and Texas mate and produce fertile progeny only in seasons of extreme climate conditions

D. Two fruit flies on the island of Bali with distinct courtship behaviors

E. All the above

16. What is the reason sponges are classified as animals?

A. Sponges form partnerships with photosynthetic organisms

B. Sponges are autotrophic

C. Their cells have cell walls

D. Sponges are heterotrophic

E. Sponges are omnivores

17. What is a protostome?

    A. outermost germ layer

    B. animal whose mouth is formed from the blastopore

    C. animal whose anus is formed from the blastopore

    D. embryo just after fertilization

    E. innermost germ layer

18. Arthropods have:

    A. spiny skin and an internal skeleton

    B. flattened body with an external shell

    C. segmented body and an exoskeleton

    D. soft body and an internal shell

    E. segmented body and an endoskeleton

19. A medical student examines a tissue slice on an unlabeled microscope slide. The student concludes the tissue is not from an animal because the cells in the tissue have:

    A. flagella

    B. cell membranes

    C. nuclei

    D. membrane-bound organelles

    E. cell walls

20. Which is true about annelids?

    A. Each body segment contains several pairs of antennae

    B. Annelids rely on diffusion to transport oxygen and nutrients to their tissues

    C. Annelids have segmented bodies and a true coelom

    D. Annelids are acoelomates

    E. Annelids have segmented bodies and a pseudocoelom

21. Ancient chordates are thought to be most closely related to:

    A. octopi

    B. sea anemones

    C. spiders

    D. earthworms

    E. none of the above

22. Which is true about echinoderms?

   A. Echinoderms have one pair of antennae and unbranched appendages

   B. Adult echinoderms have an exoskeleton

   C. Most adult echinoderms exhibit radial symmetry

   D. Echinoderm's body is divided into a head, thorax, and abdomen

   E. Most adult echinoderms lack radial symmetry

23. The simplest animals to have body symmetry are:

   A. cnidarians

   B. echinoderms

   C. sponges

   D. algae

   E. bacteria

24. In chordates whose pharyngeal pouches develop slits for breathing, leading outside the body, adults use:

   A. pharynx

   B. nose

   C. gills

   D. lungs

   E. larynx

25. In a chordate embryo, nerves branch in intervals from:

   A. spinal column

   B. notochord

   C. pharyngeal pouches

   D. tail

   E. hollow nerve cord

26. Each is an essential function performed by animals, EXCEPT:

   A. circulation

   B. cephalization

   C. excretion

   D. respiration

   E. all are required

27. Which animals are deuterostomes?

   A. mollusks and arthropods

   B. arthropods and chordates

   C. annelids and arthropods

   D. cnidarians and mollusks

   E. echinoderms and chordates

28. In fishes, pharyngeal pouches may develop into:

A. gills

B. tails

C. lungs

D. fins

E. larynx

29. What is true about corals?

A. The polyp form of corals is restricted to a small larval stage

B. Corals only reproduce by asexual means

C. Corals have only the medusa stage in their life cycles

D. Coral polyps secrete an underlying skeleton of calcium carbonate

E. One coral polyp forms a balloon-like float that keeps the entire colony afloat

30. Amphibians evolved from:

A. jawless fishes

B. lobe-finned fishes

C. ray-finned fishes

D. cartilaginous fishes

E. none of the above

31. The notochord is responsible for which function in an embryo?

A. respiration

B. processing nerve signals

C. processing wastes

D. cognitive abilities

E. structural support

32. A flexible, supporting structure found only in chordates is the:

A. pharyngeal slits

B. dorsal fin

C. nerve net

D. notochord

E. vertebrates

33. Which is NOT true about earthworms?

A. Earthworms are hermaphrodites that reproduce sexually

B. Earthworms have a digestive tract that includes a mouth and an anus

C. Earthworms have a pseudocoelom

D. Their tunnels allow for the growth of oxygen-requiring soil bacteria

E. Earthworms feed on live and dead organic matter

34. Skeletons of early vertebrates were composed of cartilage instead of bone. Which characteristic does cartilage share with notochords?

A. Oxygen can diffuse through it

B. It contracts

C. It is soft and flexible

D. It is hard and rigid

E. It contains hydroxyapatite

35. Which do NOT exhibit bilateral symmetry?

A. arthropods

B. cnidarians

C. mollusks

D. annelids

E. primates

36. Which chordate characteristic is visible on the outside of an adult cat?

A. a tail that extends beyond the anus

B. pharyngeal pouches

C. hollow nerve cord

D. notochord

E. nerve plexus

37. Fewer than 5 percent of animal species have:

A. cell membranes

B. vertebral columns

C. a protostome development pattern

D. eukaryotic cells

E. membrane-bound organelles

38. The eggs of ovoviviparous fish are:

A. released from the female before they are fertilized

B. nourished by a structure similar to a placenta

C. released from the female immediately after being fertilized

D. retained and nourished by the female

E. held inside the female's body as they develop

39. Which statement about chordates is true?

A. All chordates are vertebrates

B. All chordates have paired appendages

C. All chordates have a notochord

D. All chordates have backbones

E. Chordates include protostomes and deuterostomes

40. Which pairs of modern chordate groups are most closely related?

A. birds and crocodilians

B. sharks and the coelacanth

C. hagfishes and lungfishes

D. lampreys and ray-finned fishes

E. crocodilians and sharks

41. The hominoid group of primates includes:

A. anthropoids only World monkeys

B. apes and hominines

C. New World monkeys and Old

D. lemurs, lorises and anthropoids

E. apes and anthropoids

42. What fossil evidence indicates the order for these three vertebrates: a bony fish with a jaw, a jawless fish, and a fish with leglike fins?

A. The jawless fish was the last to evolve

B. The fish with leglike fins evolved before the jawless fish

C. The bony fish evolved before the jawless fish

D. The fish with leglike fins was the last to evolve

E. The bony fish was the last to evolve

43. Jaws and limbs are characteristic of:

A. reptiles only

B. invertebrates only

C. all chordates

D. worms

E. vertebrates only

44. Which structure in the amniotic egg provides nutrients for the embryo as it grows?

A. yolk

B. allantois

C. chorion

D. amnion

E. shell

45. Which trend in reproduction is evident when following vertebrate groups?

A. progressively fewer openings in the digestive tract

B. closed circulatory system becomes open

C. internal fertilization evolves into external fertilization

D. endothermy evolves into ectothermy

E. external fertilization evolves into internal fertilization

46. A tail that is adapted for grasping and holding objects is:

A. prehensile

B. radial

C. bipedal

D. binocular

E. symmetrical

47. Which is NOT a characteristic of mammals?

A. hair

B. endothermy

C. ability to nourish their young with milk

D. lack of pharyngeal pouches during development

E. three middle ear bones

48. The group that includes gibbons and humans but does not include tarsiers is the:

A. mollusks

B. hominines

C. primates

D. anthropoids

E. hominoids

49. All are examples of cartilaginous fishes EXCEPT:

A. rays

B. sharks

C. hagfishes

D. skates

E. sturgeons

50. Old World monkeys can be distinguished from New World monkeys by observing:

A. how the monkeys use their tails

B. when the monkeys are most active

C. what the monkeys eat

D. how the monkeys interact with their troop

E. how the monkeys gather food

51. Compared with that of a reptile, a bird's body temperature is:

A. lower and more constant

B. lower and more variable

C. higher and more constant

D. higher and more variable

E. the same

52. Which internal characteristic would an earthworm have?

A. prostomium

B. notochord

C. pseudocoelom

D. paired organs

E. blastopore

53. Mammals that lay eggs are called:

A. reptiles

B. placentals

C. marsupials

D. amphibians

E. monotremes

54. The presence of legs or other limbs indicates that the animal is:

A. protostome

B. segmented

C. acoelomate

D. radially symmetrical

E. deuterostome

55. A primatologist finds a new primate species in a Madagascar forest. The primate has a long snout and is active at night. A primatologist would classify this primate as a:

A. bush baby

B. hominoid

C. anthropoid

D. lemur

E. none of the above

56. The vertebrate group with the most complex respiratory system is:

A. fish

B. amphibians

C. mammals

D. reptiles

E. birds

57. Having a thumb that can move against the other fingers makes it possible for a primate to:

A. judge the locations of tree branches

B. display elaborate social behaviors

C. merge visual images

D. hold objects firmly

E. develop elaborate social communications

58. Which is NOT a characteristic of chordates?

A. pharyngeal pouches

B. vertebrae

C. tail that extends past the anus

D. dorsal, hollow nerve cord

E. none of the above

59. An animal that has body parts that extend outward from its center shows:

A. bilateral symmetry

B. radial symmetry

C. segmentation

D. several planes of symmetry

E. protostome development

60. Which is a node expected on the cladogram for animals?

A. leg length

B. feather arrangement

C. deuterostome development

D. fur texture

E. shape of forelimbs

61. Which variation is expected for land vertebrates?

A. Shapes of forelimbs

B. Numbers of germ layers

C. Numbers of limbs

D. Types of symmetry

E. Anteroposterior axis

62. The earliest hominine that belonged to the same genus as modern humans were probably:

A. *Australopithecus afarensis*

B. *Homo ergaster*

C. *Homo sapiens*

D. *Homo neanderthalensis*

E. *Homo habilis*

63. The most important characteristic that separates birds from other chordates is:

A. amniotic egg

B. feathers

C. wings

D. four-chambered heart

E. none of the above

64. A double-loop circulatory system is present in all animals with:

A. two-chambered heart

B. lungs

C. jaws

D. vertebrae

E. bilateral symmetry

65. Which statement about the cladogram of animals is true?

    A. Segmentation evolved more than once in different branches of the cladogram

    B. All deuterostomes have cephalization

    C. Backbones evolved before segmentation

    D. Radial symmetry appears only once in the cladogram

    E. All deuterostomes have bilateral symmetry

66. All are reasons why green algae are classified as plants, EXCEPT:

    A. They have genes similar to genes in land plants

    B. They have chlorophyll *a*

    C. They have cellulose in their cell walls

    D. They are single-celled organisms

    E. They have chlorophyll *b*

67. Which statement about green algae is true?

    A. Like other plants, they have specialized structures

    B. They are multicellular plants

    C. Evidence suggests they were the first plants

    D. They live in dry areas on land

    E. none of the above

Notes or active learning

Notes or active learning

Detailed Explanations: Classification & Diversity

Answer Key

1: D	11: B	21: B	31: E	41: B	51: C	61: A
2: A	12: A	22: C	32: D	42: D	52: D	62: E
3: E	13: A	23: A	33: C	43: E	53: E	63: B
4: A	14: E	24: C	34: C	44: A	54: B	64: B
5: C	15: B	25: E	35: B	45: E	55: D	65: A
6: B	16: D	26: B	36: A	46: A	56: E	66: D
7: B	17: B	27: E	37: B	47: D	57: D	67: C
8: E	18: C	28: A	38: E	48: E	58: B	
9: B	19: E	29: D	39: C	49: C	59: B	
10: D	20: C	30: B	40: A	50: A	60: C	

1. D is correct.

Mnemonic: **D**arn **K**ing **P**hillip **C**ame **O**ver **F**or **G**ood **S**oup:

> Domain → Kingdom → Phylum → Class → Order
> → Family → Genus → Species

Canis lupus indicates that the wolf's genus is *Canis,* and the species is *lupus.*

Family is more inclusive than genus or species, so any member of the species *lupus* is in the genus *Canis* of the Canidae family.

2. A is correct.

Class encompasses several orders.

Mnemonic: **D**arn **K**ing **P**hillip **C**ame **O**ver **F**or **G**ood **S**oup:

> Domain → kingdom → phylum → class → **order** → family → genus → species

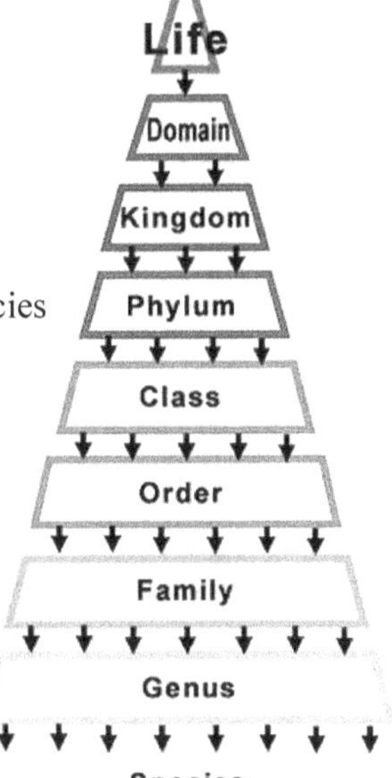

3. E is correct.

Homology serves as evidence of common ancestry.

Homology is a relationship between structures or DNA derived from a common ancestor.

Homologous traits are traced by descent from a common ancestor.

continued...

Analogous organs have similar functions in two taxa, not in the last common ancestor but evolved separately.

Homologous sequences are *orthologous* if they descend from the same ancestral sequence separated by a speciation event.

When a species diverges into two species, the copies of a single gene in the resulting species are orthologous.

Orthology is strictly defined by ancestry.

Orthologs (i.e., orthologous genes) are genes in distinct species that originated by vertical descent from a single gene of the last common ancestor.

Due to gene duplication and genome rearrangement, the ancestry of genes is difficult to ascertain.

Strongest evidence that two similar genes are orthologous is usually a phylogenetic analysis of the gene lineage.

Orthologous genes often, but not always, have the same function.

4. A is correct.

Homology is a relationship between structures or DNA derived from a common ancestor.

Homologous traits descend from a common ancestor (e.g., forelimbs of a human and a dog).

Analogous organs, the opposite of homologous organs, have similar functions in two taxa, not in the last common ancestor but evolved separately.

Homologous sequences are *orthologous* if they descend from the same ancestral sequence separated by a speciation event.

When a species diverges into two separate species, the copies of a single gene in the resulting species are orthologous.

Orthologues (i.e., orthologous genes) are genes in distinct species that originated by *vertical descent* from a single gene of the last common ancestor.

Orthology is defined by ancestry.

With gene duplication and genome rearrangement, genetic ancestry is difficult to ascertain.

Strongest evidence that two similar genes are orthologous is through phylogenetic analysis of gene lineage.

Orthologous genes often, but not always, have the same function.

continued...

	Analogous leg	Analogous flipper
Homologous mammals	Cat leg	Whale flipper
Homologous insects	Praying mantis leg	Water boatman flipper

Homologous (common ancestor) vs. analogous structures (similar function but evolved separately)

5. C is correct.

Homology is a relationship between structures or DNA derived from a common ancestor.

Homologous traits descend from a common ancestor (e.g., forelimb of a human and a dog).

Analogous organs are the opposite of homologous organs and have similar functions in two taxa that were not in the last common ancestor but evolved separately.

Homologous sequences are *orthologous* if they descend from the same ancestral sequence separated by a speciation event.

When a species diverges into two species, copies of a single gene in the two resulting species are orthologous.

Orthologs (i.e., orthologous genes) are genes in distinct species that originated by vertical descent from a single gene of the last common ancestor.

Orthology is strictly defined by ancestry.

Strongest evidence that two similar genes are orthologous is usually a phylogenetic analysis of the gene lineage.

Orthologous genes often, but not always, have the same function.

D: wings of a pigeon and a bat represent *analogous structures*.

6. B is correct.

Taxonomy of humans:

> Domain: Eukaryota → Kingdom: Animalia → Phylum: Chordata → Subphylum: Vertebrata →
> Class: Mammalia → Order: Primates → Family: Hominidae → Genus: Homo →
> Species: *H. sapiens* → Subspecies: *H. s. sapiens*

7. B is correct.

Ants are not chordates.

Tunicates are marine invertebrates and members of the subphylum Tunicata within Chordata (i.e., the phylum includes all animals with dorsal nerve cords and notochords).

Tunicates live as solitary individuals or replicate by budding and becoming colonies (each unit is a zooid).

Tunicates are marine filter feeders with a water-filled, sac-like body structure and two tubular openings, known as *siphons*, through which they draw in and expel water.

They take in water through the *incurrent siphon* during respiration and feeding and expel filtered water through the *excurrent siphon*.

Most adult tunicates are sessile and permanently attached to rocks or other hard surfaces on the ocean floor; others swim in the pelagic zone (i.e., open sea) as adults.

8. E is correct.

Echinoderms are invertebrate predecessors of the chordates and include starfish, sea urchins, and cucumbers.

They are characterized by a primitive vascular system known as the water vascular system.

Adult echinoderms have radial symmetry, while larvae are bilaterally symmetrical.

They move with structures known as tube feet, have no backbone, and are heterotrophic (i.e., they do not synthesize food).

Crayfish are in the phylum Arthropoda, which includes insects.

Arthropods are characterized by segmented bodies covered in a chitin exoskeleton and jointed appendages.

9. B is correct.

Kingdom encompasses many phyla.

Mnemonic: **D**arn **K**ing **P**hillip **C**ame **O**ver **F**or **G**ood **S**oup:

Domain → kingdom → **phylum**
→ class → order → family →
genus → species

10. D is correct.

Homo sapiens belong to the phylum Chordata.

Taxonomy of humans:

Domain: Eukaryota → Kingdom: Animalia
→ Phylum: Chordata → Subphylum: Vertebrata →

Class: Mammalia → Order: Primates
→ Family: Hominidae → Genus: Homo
→ Species: *H. sapiens* → Subspecies: *H. s. sapiens*

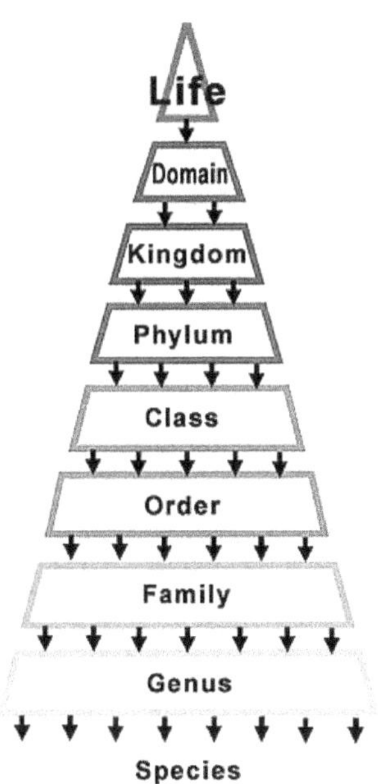

11. B is correct.

Organisms of the phylum *Chordata* have a dorsal notochord, while subphylum Vertebrata has a backbone.

Nonvertebrate chordates include amphioxus (lancelet) or unrelated tunicate worm.

Lancelet (or *amphioxus*) is the modern subphylum Cephalochordata and is vital in zoology – it provides indications about the origins of the vertebrates.

Lancelets are comparison points for tracing vertebrates' evolution.

They are the archetypal vertebrate form.

Lancelets split from vertebrates more than 520 million years ago; their genomes hold clues about evolution, specifically how vertebrates have adapted old genes for new functions.

Vertebrates include most of the phylum Chordata with about 64,000 species.

Vertebrates include amphibians, reptiles (e.g., lizards), mammals, birds, jawless and bony fish, sharks, and rays.

Shark and lamprey eel are vertebrates with cartilaginous skeletons.

12. A is correct.

Cell, tissue, organ, organism, population, and community represent the correct ordering of the levels of complexity at which life is studied, from simplest to complex.

13. A is correct.

Notochord is *ventra*l to the *neural tube* and forms during gastrulation.

The notochord induces *neural plate formation* (during neurulation) to synchronize neural tube development (precursor to the central nervous system).

Notochord is a *flexible rod-shaped structure* in embryos of all chordates.

Notochord comprises cells derived from *mesoderm* and defines the *primitive axis* of the embryo.

In lower chordates, this chord remains throughout the life of the animal.

In higher chordates (e.g., humans), the notochord exists during embryonic development and disappears.

If the notochord persists, it functions as the primary axial support. In most vertebrates, it becomes the nucleus pulposus of the intervertebral disc.

B: *notochord remains* in the lower chordates, such as the amphioxus and the tunicate worm.

It is not vestigial (not a genetically determined structure that lost its ancestral function) because it disappears.

D: *echinoderms* are the invertebrate predecessors of chordates and include starfish, sea urchins, and cucumber.

Echinoderms do *not* possess a notochord.

E: *notochord is not* part of the nervous system.

14. E is correct.

Chordates are animals that, for at least some of their life cycle, possess:

      1) a notochord

      2) a dorsal neural tube

      3) pharyngeal slits

      4) an endostyle

      5) a post-anal tail

continued…

Chordate phylum includes the subphyla Vertebrata (e.g., mammals, fish, amphibians, reptiles, birds),

Tunicata (e.g., salps, sea squirts), and Cephalochordata (e.g., lancelets).

Vertebrata subphylums have a backbone, but it is not a shared characteristic of chordates.

Chordates are a phylum that shares a bilateral body plan and have, at some stage:

> *Notochord* is a relatively stiff rod of cartilage extending inside the body.

> Among vertebrate sub-group chordates, the notochord develops into the spine.

> This helps the animal swim by flexing its tail in aquatic species.

> *Dorsal neural tube* (for vertebrates, including fish) develops into the spinal cord, the leading communications trunk of the nervous system.

> *Pharyngeal slits* comprise the throat immediately behind the mouth (modified fish gills).

> In some chordates, pharyngeal slits are part of a filter-feeding system extracting food from its environment.

> *Endostyle groove* in the ventral wall of the pharynx stores iodine and a precursor of the vertebrate thyroid gland.

> In filter-feeders, it produces mucus to gather food particles, transporting food to the esophagus.

> *Post-anal tail* is a muscular tail extending behind the anus.

15. B is correct.

Migratory birds nesting on islands is not behavioral, temporal, or geographical isolation leading to speciation.

The birds are migratory and are not geographically isolated.

16. D is correct.

Sponges are Porifera (i.e., pore or Ostia-bearing) phylum animals.

Like other animals, they are multicellular, heterotrophic, and lack cell walls.

Sponges lack tissues and organs and have no body symmetry, differentiating them from other animals.

Heterotrophy means obtaining food and energy by consuming other organic substances rather than sunlight or inorganic compounds.

17. B is correct.

Homology is a relationship between structures or DNA derived from a common ancestor.

Homologous traits are explained by descent from a common ancestor.

Analogous organs have similar functions in two taxa, not in the last common ancestor but evolved separately.

Homologous sequences are *orthologous* when descended from ancestral sequences separated by speciation. When a species diverges into two species, copies of a single gene in the two resulting species are *orthologous*.

Orthologues (i.e., orthologous genes) are genes in distinct species that originated by vertical descent from a single gene of the last common ancestor.

Orthology is defined by ancestry.

Due to gene duplication and genome rearrangement, the ancestry of genes is difficult to ascertain.

Strongest evidence that two similar genes are orthologous is usually a *phylogenetic analysis* of *gene lineage.*

Orthologous genes often, but not always, have the same function.

	Analogous leg	Analogous flipper
Homologous mammals	Cat leg	Whale flipper
Homologous insects	Praying mantis leg	Water boatman flipper

Homologous (common ancestor) vs. analogous structures (similar function but evolved separately)

18. C is correct.

Arthropods are invertebrate animals with an exoskeleton, segmented body, and jointed appendages and include insects, arachnids, myriapods, and crustaceans.

19. E is correct.

Animal cells do *not* have cell walls, distinct from plant cells.

Another distinct feature of plant cells is the presence of chloroplasts in the structure.

20. C is correct.

Annelids are a large phylum of segmented worms, each segment with the same set of organs.

Coelom is a cavity lined by a mesoderm-derived epithelium.

Organs formed inside a coelom can grow, freely move, and develop independently of the body wall while protected by fluid cushions.

Organisms are classified into three groups:

1) *Coelomates* are animals with a fluid-filled body cavity (i.e., coelom) and complete lining (peritoneum derived from mesoderm), allowing organs to be attached and suspended while moving freely within the cavity.

Coelomates include most bilateral animals, including all vertebrates.

2) *Pseudocoelomates* are animals with a pseudocoelom fully functional body cavity (e.g., roundworm).

Pseudocoelomate *organs* are held loosely and not as well organized as coelomates.

Mesoderm-derived tissue partly lines the fluid-filled body cavity.

Pseudocoelomates are *protostomes*, but not all protostomes are pseudocoelomates.

3) *Acoelomates* are animals with no body cavities (i.e., flatworms).

Semi-solid mesodermal tissues between the gut and body wall hold their organs in place.

21. B is correct.

From embryological studies, ancient chordates were closely related to the ancestors of echinoderms, marine animals such as sea stars and sea urchins.

Sea anemones are a group of water-dwelling, predatory animals.

22. C is correct.

Echinoderms are marine animals such as sea stars, sea urchins, and sand dollars characterized by radial (usually five-point) symmetry.

Most echinoderms can regenerate tissue, organs, and limbs and reproduce asexually.

23. A is correct.

Cnidaria is a phylum of animals with over 10,000 species exclusively in aquatic and marine environments.

Cnidaria bodies consist of a *non-living jelly-like substance* sandwiched between two layers of epithelium that are mostly one cell thick.

Their distinguishing feature is *cnidocytes*, specialized cells mainly for capturing prey.

Cnidaria has two basic body forms: *swimming medusae* and *sessile polyps*; each is radially symmetrical, with mouths surrounded by tentacles that bear cnidocytes.

Both forms have a single orifice and body cavity for digestion and respiration.

Many cnidarian species form colonies of single organisms with Medusa-like or polyp-like zooids or both.

B: *echinoderms* are a phylum of marine animals, and adults have radial (usually five-point) symmetry (e.g., sea urchins, sand dollars, starfish, and sea cucumbers).

24. C is correct.

Chordates are animals with:

> *notochord* (stiff rod of cartilage that extends inside the body),
>
> *dorsal neural tube*, and
>
> *pharyngeal slits* are openings of the pharynx to allow aquatic organisms to extract oxygen from water and excrete carbon dioxide (i.e., gills).

25. E is correct.

A hollow *nerve cord* (a single hollow nerve tissue tract) constitutes the central nervous system of chordates.

26. B is correct.

Cephalization is an evolutionary trend when nervous tissue becomes concentrated toward one end of an organism over many generations, producing a head region with sensory organs.

continued...

Cephalization is intrinsically connected with a change in symmetry.

Cephalization accompanied the shift to bilateral symmetry in flatworms.

In addition to a concentration of sense organs, annelids place the mouth in the head region.

This process is tied to developing an anterior brain in the chordates from the notochord.

An exception to the trend of cephalization throughout evolutionary advancement is the phylum Echinodermata.

Echinodermata, despite having a bilateral ancestor, developed with 5-point symmetry) and no concentrated neural ganglia or sensory head region.

Some echinoderms have developed bilateral symmetry secondarily.

27. E is correct.

Deuterostomes differ from protostomes in embryonic development.

In deuterostomes, the first opening (i.e., blastopore) becomes the anus. In protostomes, it becomes the mouth.

Chordates, echinoderms, and hemichordates are deuterostomes.

28. A is correct.

Pharyngeal pouches are filter-feeding organs in fish that develop into gills used for respiration.

29. D is correct.

Corals are marine invertebrates (phylum Cnidaria), typically in compact colonies with identical individual *polyps*.

Coral includes reef builders of tropical oceans secreting calcium carbonate, forming a hard skeleton.

A coral "head" is a colony of myriad genetically identical polyps.

Each polyp is a spineless animal, typically a few millimeters in diameter and a few centimeters in length.

Tentacles surround a central mouth opening, and an *exoskeleton* is excreted near the base.

Over many generations, colonies create a large skeleton structure.

Individual heads grow by the asexual reproduction of polyps.

Corals breed *sexually* by *spawning:* polyps of the same species release gametes simultaneously for several nights, cycling around full moons.

30. B is correct.

First amphibians developed from lobe-finned fish about 370 million years ago during the Devonian period.

Lobe-finned fish had multi-jointed leg-like fins that allowed crawling at the bottom of the sea.

Their bony fins evolved into limbs, ancestors of tetrapods, including amphibians, reptiles, birds, and mammals.

31. E is correct.

Notochord is a flexible rod essential in vertebrate development and skeletal elements of developing embryos.

Notochord is related to cartilage, serving as the axial skeleton of the embryo until other elements form.

32. D is correct.

Chordates are animals that, for at least some period of their life cycles, possess the following: a notochord, a dorsal neural tube, pharyngeal slits, an endostyle, and a post-anal tail.

Notochord is a flexible rod-shaped structure in embryos of chordates. It is a relatively stiff cartilage rod extending along the inside of the body.

It is composed of mesoderm-derived cells that define the embryo axis.

Among the vertebrate sub-group of chordates, the notochord develops into the spine, and in aquatic species, this helps the animal swim by flexing its tail.

In lower chordates, the notochord remains throughout life.

In higher chordates (e.g., humans), the notochord exists only during embryonic development and disappears.

If it persists throughout life, it functions as the primary axial support, while in most vertebrates, it becomes the nucleus pulposus of the intervertebral disc.

Notochord is ventral to the neural tube and forms during gastrulation.

The notochord *induces neural plate formation* (during neurulation) to synchronize neural tube development (precursor to the central nervous system).

33. C is correct.

Earthworms are part of the Annelid phylum.

Pseudocoelom is a fluid-filled cavity between the body wall and intestine of certain invertebrates, such as roundworms and hookworms.

All other statements are true.

34. C is correct.

Notochords are similar to cartilage, made of type II collagen, making it soft and flexible.

Notochord is a flexible rod between the nerve cord and the digestive tract.

35. B is correct.

Cnidaria bodies consist of a non-living jelly-like substance sandwiched between two layers of epithelium that are mostly one cell thick.

Cnidaria has two body forms: swimming medusae and sessile polyps, each *radially symmetrical* with mouths surrounded by tentacles with cnidocytes.

Echinoderms are a phylum of marine animals, and adults have radial (usually five-point) symmetry (e.g., sea urchins, sand dollars, starfish, and sea cucumbers).

36. A is correct.

Chordates share several key features, such as a notochord, a dorsal hollow nerve cord, pharyngeal slits, and a post-anal tail.

Of the characteristics visible on a cat, a tail extending beyond the anus makes it a chordate.

37. B is correct.

Vertebral column is the backbone or spine, a segmented series of bones separated by intervertebral discs.

38. E is correct.

Ovoviviparous fish (e.g., guppies and angel sharks) have eggs that develop inside the mother's body after *internal fertilization*, whereby each embryo develops within its egg.

Yolk nourishes the embryo, which receives little or no nourishment from the mother.

Viviparous fish retain the eggs and nourish the embryos. They have a structure analogous to the placenta (mammals), connecting the mother's blood with the embryo.

39. C is correct.

Chordates are defined by the presence of a *notochord*.

Not all chordates are vertebrates, have paired appendages, or have backbones.

Chordates are *deuterostomes* where the first opening (i.e., blastopore) becomes the anus.

Chordates are a phylum that shares a bilateral body plan and have, at some stage:

Notochord is a relatively stiff rod of cartilage extending inside the body.

Among vertebrate sub-group chordates, the notochord develops into the spine. This helps the animal swim by flexing its tail in aquatic species.

Dorsal neural tube (for vertebrates, including fish) develops into the spinal cord, the leading communications trunk of the nervous system.

Pharyngeal slits comprise the throat immediately behind the mouth (modified fish gills). In some chordates, they are part of a filter-feeding system extracting food from the environment.

Endostyle groove in the ventral wall of the pharynx stores iodine and a possible precursor of the vertebrate thyroid gland.

In filter-feeders, it produces mucus to gather food particles, which help transport food to the esophagus.

Post-anal tail is a muscular tail extending behind the anus.

40. A is correct.

Chordates are animals with a notochord, a dorsal neural tube, pharyngeal slits, an endostyle, and a post-anal tail for at least some of their life cycles.

41. B is correct.

Hominoids are called *apes*, but *ape* is used differently.

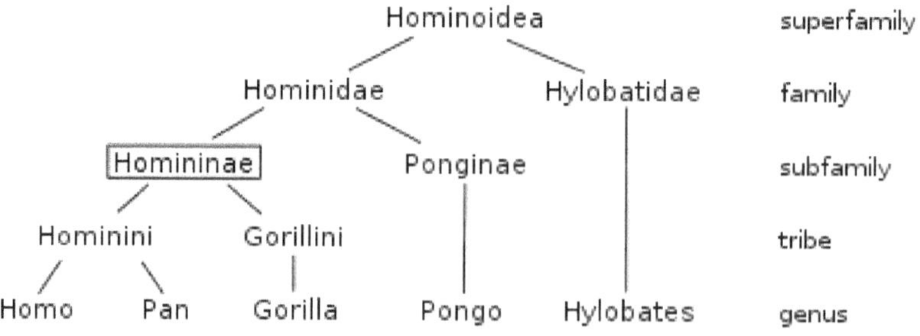

Homininae cladogram divides into genus homo, pan and gorilla

continued...

Until 1970, Hominidae meant humans only, with non-human great apes assigned to the family Pongidae.

Later discoveries led to revised classifications, with the great apes then united with humans (now in subfamily Homininae) as members of the family Hominidae

By 1990, molecular biology techniques classified gorillas and chimpanzees as more closely related to humans than orangutans.

Gorillas and chimpanzees are in the subfamily Homininae.

Hominines are a subfamily of Hominidae that includes humans, gorillas, chimpanzees, bonobos, and hominids which arose after the evolutionary split from orangutans.

Homininae cladogram has three main branches, which lead to gorillas, chimpanzees, bonobos, and humans.

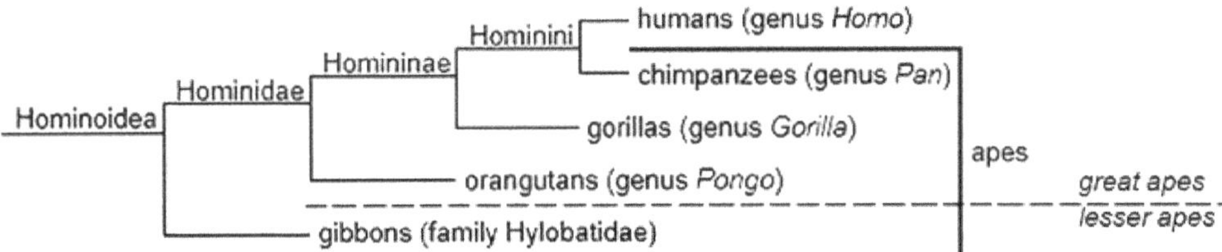

Homininae cladogram includes gorillas, chimpanzees, and humans

42. D is correct.

Fish evolved with *jawless fish*, followed by *bony fish with a jaw*.

Lastly, fish with leg-like fins, which allowed for movement along the seabed, evolved into limbs of *tetrapods*.

43. E is correct.

Jaws and limbs are characteristic of *vertebrates*.

44. A is correct.

Yolk nourishes the embryo, which receives little or no nourishment from the mother.

Amniotic fluid, surrounded by the amnion, is a liquid environment around the egg, protecting it from shock.

Embryonic membranes include:

      1) ***chorion*** lines the inside of the shell and permits gas exchange;

continued...

---

2) ***allantois*** is a saclike structure developed from the digestive tract and functions in respiration, excretion, and gas exchange with the external environment;

3) ***amnion*** encloses *amniotic fluid* as a watery environment for embryogenesis and shock protection;

4) ***yolk sac*** encloses the yolk and transfers food to the developing embryo.

45. E is correct.

External fertilization is male sperm fertilizing an egg outside the female's body, as seen in amphibians and fish.

Further in the evolution of vertebrates, mammals use internal fertilization for reproduction.

46. A is correct.

Prehensile refers to an appendage or organ adapted for holding or grasping.

47. D is correct.

Mammals are *endothermic* (warm-blooded) vertebrates with *body hair* and the ability to feed babies with *milk*.

In the embryonic development of vertebrates (e.g., mammals), *pharyngeal pouches* develop into essential structures such as the eardrum and thymus gland.

48. E is correct.

Hominoids are called *apes*, but *ape* is used differently.

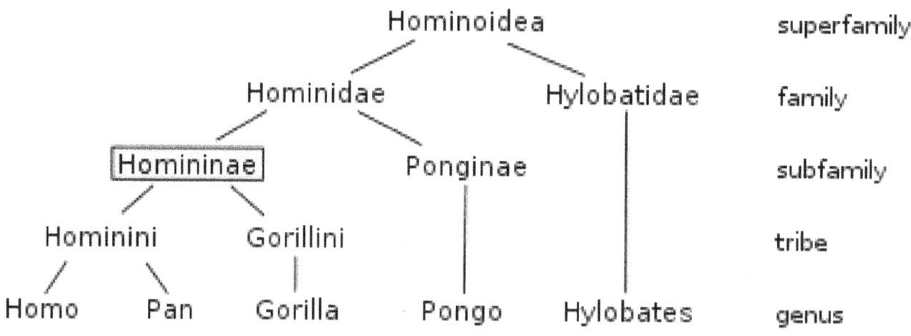

Homininae cladogram divides into genus homo, pan and gorilla

Until 1970, Hominidae meant humans only, with non-human great apes assigned to the family Pongidae.

continued...

Later discoveries led to revised classifications, with the great apes then united with humans (now in subfamily Homininae) as members of the family Hominidae

By 1990, molecular biology classified gorillas and chimpanzees as more closely related to humans than orangutans.

Gorillas and chimpanzees are in the subfamily Homininae.

Hominines are a subfamily of Hominidae that includes humans, gorillas, chimpanzees, bonobos, and hominids which arose after the evolutionary split from orangutans.

Homininae cladogram has three main branches, which lead to gorillas, chimpanzees, bonobos, and humans.

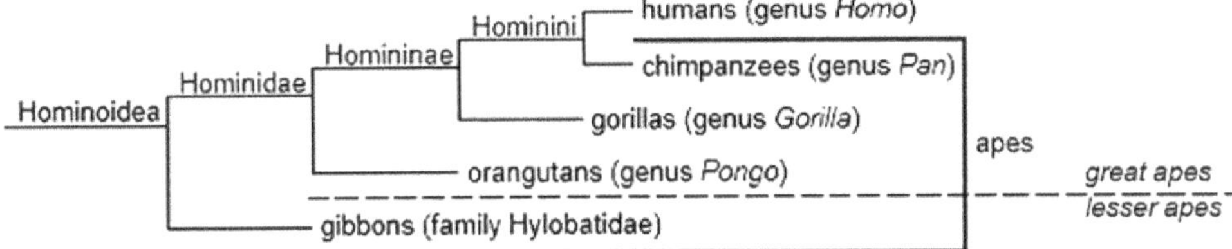

Homininae cladogram includes gorillas, chimpanzees, and humans

49. C is correct.

Cartilaginous fishes have a skeleton made of cartilage rather than bone.

Sharks, skates, rays, and sturgeons have cartilaginous skeletons, while hagfish have cartilaginous skulls.

50. A is correct.

Old World monkeys are primates in the clade of Catarrhini.

They are native to Africa and Asia today, inhabiting environments from tropical rainforest to savanna, shrubland, and mountainous terrain.

Old World monkeys include familiar species of nonhuman primates (e.g., baboons and macaques).

They are medium to large and range from arboreal (i.e., locomotion in trees) to fully terrestrial (e.g., baboons).

New World monkeys are the five families of primates in Central and South America and portions of Mexico. They are small to mid-sized primates.

New World monkeys differ from Old World monkeys in several aspects.

continued...

The prominent phenotypic distinction is the nose, commonly used to distinguish between groups.

New World monkeys have noses flatter than the narrow noses of Old-World monkeys and have side-facing nostrils.

New World monkeys are the only monkeys with prehensile tails.

Old-World monkeys have shorter, non-grasping tails.

51. C is correct.

A reptile's body temperature is determined by the temperature of its surrounding environment, meaning it could fluctuate more often than endothermic (*warm-blooded*) animals.

Birds are *endothermic*, so their body temperature remains constant and is warmer than reptiles.

52. D is correct.

Earthworms are soft-bodied segmented worms with nephridia, paired organs in each worm segment.

Nephridia remove nitrogenous waste.

53. E is correct.

Monotremes are mammals laying eggs instead of birthing live young, like marsupials and placental mammals.

The surviving species of monotremes are indigenous to Australia and New Guinea, though evidence suggests they were once widespread.

The existing monotreme species are the *platypus* and four species of echidnas (i.e., spiny anteaters).

54. B is correct.

Legs and limbs indicate that the animal is *segmented.*

55. D is correct.

Lemurs are a clade of *strepsirrhine primates* characterized by a *moist nose tip.*

Lemurs are primarily *nocturnal animals* and are on the island of Madagascar.

56. E is correct.

Birds have a *complex respiratory system.*

Upon inhalation, 75% of the fresh air bypasses the lungs and flows directly into a posterior air sac extending from the lungs, connects with air spaces in the bones, and fills them with air.

25% of the air goes directly into the lungs.

When birds exhale, the used air flows out of the lung, and the stored fresh air from the posterior air sac is simultaneously forced into the lungs.

Bird's lungs receive fresh air during inhalation and exhalation.

Sound production uses the *syrinx*, a muscular chamber incorporating multiple tympanic membranes that diverge from the trachea's lower end.

In some species, the *trachea* is elongated to increase *vocalization volume* and *perceived size.*

57. D is correct.

Grasping hands of primates (i.e., having opposable thumbs) are an adaptation to life in the trees.

Thumbs help primates grasp objects (e.g., tree branches) firmly.

58. B is correct.

Chordates are both vertebrates and invertebrates.

Chordates share features:

> *notochord,*
>
> *dorsal hollow nerve cord,*
>
> *pharyngeal slits*, and
>
> *post-anal tail.*

In vertebrate chordates, the notochord is replaced by a *vertebral column.*

59. B is correct.

Radial symmetry is present in organisms with symmetry around a central axis.

Radial symmetry is in *starfish* or *sea urchins*, whose body parts extend outward from a center.

60. C is correct.

Deuterostome is distinguished by embryonic development, whereby the first opening (i.e., blastopore) becomes the anus.

Protostomes develop with the first opening (i.e., blastopore) as the *mouth*.

Deuterostome mouth develops at the opposite end of the embryo from the blastopore. A digestive tract develops in the middle connecting the two.

Humans are deuterostomes.

Invertebrates are *protostomes.*

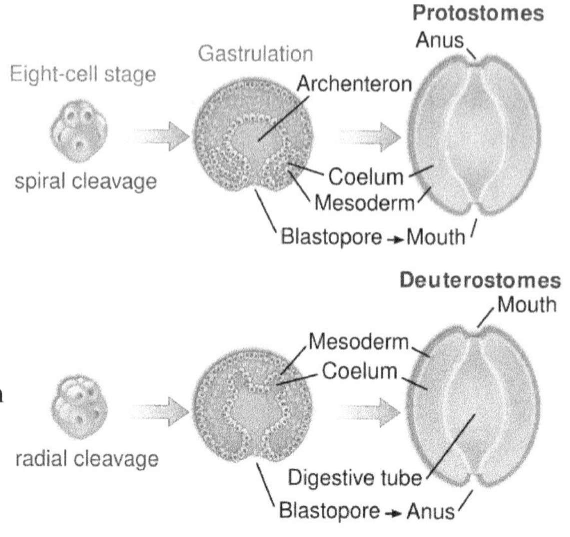

Protostomes (mouth at blastopore) vs. deuterostomes

61. A is correct.

Shapes of forelimbs are an expected variant for vertebrates using these appendages for utility.

62. E is correct.

Homo habilis (or *Australopithecus habilis*) is a species of the *Hominini* tribe, which lived from 2.33 to 1.44 million years ago.

There has been controversy regarding the placement of *Homo habilis* in the genus *Homo* rather than *Australopithecus*; its brain size is 550-687 cm^3 rather than 363-600 cm^3.

Recent findings on brain size support its traditional placement in the genus *Homo* and the theory that *Homo habilis* is the common ancestor.

63. B is correct.

Feathers are the most critical characteristic separating birds from other living animals.

Birds are the only *chordates* with feathers.

Other chordates, such as mammals, have *fur or hair.*

64. B is correct.

Double-loop circulatory system is one in which blood flows through the heart twice.

Pulmonary circulation involves blood flowing between the heart and lungs.

Systemic circulation is blood movement from the heart to the body and back.

Lungs are essential for *double-loop circulatory systems*.

65. A is correct.

Cladograms of animals support segmentation evolving more than once in different branches.

Cephalization is an evolutionary trend whereby nervous tissue (over many generations) becomes concentrated toward one end of an organism to produce a head region with sensory organs.

66. D is correct.

Green algae are classified as *plants* because they have *gene segments* similar to land plants, use chlorophyll *a*, and have cellulose cell walls.

67. C is correct.

Evidence suggests that green algae were the first plants.

Land plants evolved from *filamentous green algae*, which invaded land about 410 million years ago

Land plants and green algae have *cellulose* in their cell walls and share many *photopigments for photosynthesis*.

Notes for active learning

Notes for active learning

Notes for active learning

Annotated Glossary

A

Acanthodians – a primitive group of Silurian to Permian jawed bony fishes, bearing bony spines in front of all but their caudal fins.

Albian – European stage of the uppermost Lower Cretaceous, between 107 and 95 million years ago.

Algae – photosynthetic, almost exclusively aquatic, nonvascular plants ranging from simple unicellular forms to giant kelps several feet long. They have extremely varied life cycles and first appeared in the Precambrian.

Alveolus (alveoli, plural) – the depressions or sockets in the tooth-bearing skeletal elements such as maxillae and dentaries into which the roots of the teeth are inserted. *Exceptionally well-developed in mammals and crocodilians.*

Ammonite – a coiled, chambered fossil shell of a cephalopod mollusk of the extinct order Ammonoidea.

Amniote – a member of the taxonomic group Amniota, such as reptiles, mammals, and birds, characterized by possessing the amnion, a specialized embryonic membrane.

Amoeba – a microscopic, one-celled animal consisting of a naked mass of protoplasm.

Aperture – a relatively large opening on the last-formed chamber of a foraminiferal shell.

Aptian – European stage of the Lower Cretaceous, between 114 and 107 million years ago.

Arboreal – an animal that primarily lives in trees.

Archaean [ancient] – the middle era of Precambrian time, between 3.8 and 2.5 billion years ago. Life arose on Earth during the early Archaean, as fossil bacteria in rocks about 3.5 billion years old indicate.

Astragalus (or *talus*) – bone articulating with the tibia and fibula to form the ankle joint.

Atlas – the first or *anterior-most* vertebra of a tetrapod that articulates with the skull.

Auditory bulla (bullae, plural; or *tympanic bulla*) – the bone that encloses the inner ear region of the mammalian skull and is found on the ventral surface of the posterior region. *Typically, fragile, and thus rarely preserved in fossils, but inflated and composed of solid bone in whales. For that reason, whale auditory bullae are commonly recovered as fossils.*

Autochthonous (*paleontology*) – describes a taxon or clade that evolved from its common ancestor in its present geographic region.

Axis (axes, plural) – the second vertebra in a tetrapod follows the atlas.

B

Barremian – European Lower Cretaceous stage, between 118 and 114 million years ago.

Basicranium – the posterior and ventral portion of the skull comprises the basioccipital, basisphenoid, petrosal, and auditory bulla and portions of squamosal and pterygoid bones. *The evolutionary relationships of mammals can often be deduced by careful study of the arrangement of these bones and the foramina through which blood vessels and nerves pass.*

Basicranial – relating to the basicranium.

Benthic – describes bottom-dwelling aquatic organisms.

Berriasian – European stage of the lowermost Lower Cretaceous, between 135 and 131 million years ago.

Bilophodont – a tooth in which the crown has two transverse ridges, for example, the lower molars of tapirs, usually used only in conjunction with mammalian teeth.

Bioluminescence – the production of light by living organisms.

Biostratigraphy – the branch of geology concerned with separating and differentiating rock units by studying the fossils they contain.

Bivalve – a mollusk with two shells hinged together, like the oyster, clam, mussel, or an animal with two halves to its shell, such as an *ostracode* or *brachiopod*.

Brachydont – a tooth in which the crown is relatively short, especially one in which the crown is shorter than the roots or an animal with such teeth. Usually used only in conjunction with mammalian teeth.

Bone – 1. any single component of the vertebrate skeletal system, regardless of composition. *For example, the maxilla, humerus, and ribs are all bones.* 2. a particular type of vertebrate skeletal tissue comprises a mixture of inorganic hydroxyapatite crystals and proteins.

Bony fishes – the class Osteichthyes, characterized by a skeleton composed of bone in addition to cartilage, gill covers, and an air bladder.

Browser – an animal whose diet is entirely or mainly the leaves and stems of shrubs, bushes, and trees.

Browse – food for animals with diets entirely or mainly leaves and stems of shrubs, bushes, and trees.

Buccal (or *labial*) – a tooth's lateral side (or *external*), towards the lips or cheek.

Bunodont – a tooth in which the crown has low, rounded, isolated cusps or cones not connected by ridges or lophs. Lophs run from the tongue side of the tooth to the cheek side, with one ridge on the *tooth's anterior (or front) aspect* and one on the *posterior* (or back). Used for mammalian teeth.

C

Caecilians – wormlike, almost blind, tropical amphibians of the order Apoda.

Calcaneum – bone forming the heel and often the largest tarsal bones. *Articulates primarily with the cuboid and astragalus bones of the foot and ankle.*

Calcareous – of, containing, or like calcite (calcium carbonate).

Calcareous nannofossils – fossil remains of calcareous nannoplankton.

Calcareous nannoplankton – protists that typically produce coccoliths during some phase in their life cycle.

Calcite – a common rock-forming mineral: $CaCO_3$. Calcite can be white, colorless, or pale shades of gray, yellow, and blue. It readily effervesces (bubbles) in hydrochloric acid and is the principal component of limestone.

Cambrian – the earliest period of the Paleozoic era, between 544 and 505 million years ago; named after Cambria, the Roman name for Wales, where rocks of this age were first studied.

Campanian – European stage of the Upper Cretaceous, between 84 and 72 million years ago.

Canine – the usually large, pointed, single-rooted tooth between the incisor and premolar series in the mammalian dentition.

Caniniform – any tooth having the sharp, conical appearance of a canine; usually used to describe an incisor or premolar that functionally resembles a typical canine.

Carapace – the bony dorsal shell of a *turtle* or *cingulate edentate* (e.g., armadillos).

Carboniferous – a period in the Paleozoic era that includes the Pennsylvanian and Mississippian periods and extended from 360 to 286 million years ago.

Carnassial – the specialized blade-like cheek teeth of carnivorous mammals used to cut meat. Usually consists of a pair of teeth (one upper, one lower) and is best developed in the families Felidae and Nimravidae. *The carnassial pair in the mammalian order Carnivora consists of the fourth upper premolar and the first lower molar.*

Carnivore – 1) any animal whose diet includes vertebrate flesh and tissues. 2) common name for a member of the mammalian order Carnivora, regardless of diet.

Carpal (or *carpus*) – any skeletal element that makes up the wrist, such as the unciform or trapezoid. Distally the carpals articulate with the metacarpals, proximally with the radius and ulna.

Carpometacarpus – in the limbs of birds, the metacarpals and some carpals have fused to form a single bone.

Carpus (or *carpal*) – any skeletal element of the wrist, such as the unciform or trapezoid. Distally the carpals articulate with the metacarpals, proximally with the radius and ulna.

Cartilage – a flexible, organic vertebrate tissue found in the skeletal system. Softer than bone, it is in places of the skeleton where flexibility is a greater asset than strength, such as on movable joint surfaces, the external ears, and the tip of the nose. Because of its ability to reshape itself (unlike bone), the skeletons of fetal and young juveniles consist primarily of cartilage, which is gradually replaced by bone tissue as the animal matures. Some animals, such as sharks, retain an internal skeleton made of cartilage as adults. Unless it contains calcium-rich minerals, cartilage is rarely preserved in the fossil record.

Cartilaginous fishes – have a skeleton composed chiefly of cartilage (e.g., sharks, rays). Cartilage is gristle or a firm, elastic, flexible connective tissue.

Caudal – relating to the tail, especially one of the vertebrae of the tail.

Cement – a soft, fibrous tissue of the vertebrate skeletal system. It surrounds the roots in the alveoli of mammals and some reptiles, adhering the tooth to the bone. Also found on the crown of the tooth in some hypsodont or hypselodont mammals, where it surrounds the enamel and fills in deep valleys and depressions, thus providing structural support.

Cenomanian – European stage of lowermost Upper Cretaceous, between 95 and 91 million years ago.

Cenozoic [Greek for *new life*] – an era of geologic time from the beginning of the Tertiary period (65 million years ago) to the present.

Centrum (**centra**, plural) – the spool-shaped or cylindrical body of a vertebra.

Cervical – relating to the neck, especially one of the neck vertebrae.

cf. [Latin, *confer*, meaning "*compare*"] – when placed in front of the scientific name of a taxon, it means there is some uncertainty about the identification. It may mean that the fossils are too incomplete to be identified with certainty or that the available specimens are not adequate to distinguish between two or more similar species.

Cheek tooth – any mammalian tooth found posterior to the canine, i.e., a premolar or molar.

Chordate – any member of the Chordata, a taxon of animals distinguished from other animals by having a notochord, pharyngeal gill slits, a dorsal main nerve chord, V-shaped segmented muscles, and a post-anal tail. *Some chordate character states are found in vertebrates only during embryonic development and are either lost or highly modified in adults. Vertebrates are the best-known group of chordates, but the taxon includes several marine groups, most notably the tunicates and the cephalochordates.*

Cingulum (**cingula**, plural) – a ridge or shelf found on cheek teeth of some mammals located on the outer margin of the tooth. Cusps protruding from the cingulum are called styles (or *styles* on lower teeth), such as a parastyle or metastyle. In certain mammalian groups, the cusp, referred to as the hypocone, initially evolved in the posterlingual region of the cingulum.

Clade – a group of organisms believed to have evolved from a common ancestor. A clade often refers to any taxonomic level up to the genus level where members within that group are evolutionarily distinguishable from other clades of a similar taxonomic rank. On a *cladogram*, a clade is represented by all branches following a *node*.

Cladistics – a classification method that defines taxa based on shared characteristics and arranges them using inferred evolutionary relationships so that all members of a given taxon share a closest common ancestor.

Cladogram – a branching diagram showing the phylogenetic relationships between species, genera, or other taxa.

Coccoliths – microscopic structures of varying shape and size made of calcite are secreted by calcareous nannoplankton and found in marine deposits from the Triassic period to the Recent. *Coccoliths range in size from one to thirty-five micrometers.*

Collagen – the major type of protein found in vertebrate bone and other hard skeletal tissues.

Conda – the primary cusps or hill-like structures found on lower mammalian teeth. A prefix indicates a specific cone, such as metacone or paracone. *Cone* is used with upper teeth.

Cone – the primary cusps or hill-like structures found on upper mammalian teeth. A prefix indicates a specific cone, such as metacone or paracone. *Conda* is used with lower teeth.

Coniacian – European stage of the Upper Cretaceous, between 90 and 88 million years ago.

Conspecific – of or relating to a single species. For example, if two specimens are identified as belonging to the same species, they would be described as *conspecific*; two members of a single species population are *conspecifics*.

Contemporary (or *contemporaneous*) – existing or occurring in the same period.

Convergence (*parallelism* or *parallel evolution*) – the independent acquisition of the same derived character state in two or more taxonomic groups not through a common ancestor.

Convergent evolution – the independent evolution of similar features in species of different lineages.

Co-occurrence – to appear together simultaneously or nearly simultaneously.

Coprolite – fossilized feces or dung.

Core – a cylindrical section of rock, usually 2-4 inches in diameter and up to several feet long, results from coring into the Earth. *Individual cores are brought to the surface for geologic examination or laboratory analysis.*

Cretaceous [Latin, *creta* for chalk] – the final period of the Mesozoic era, between 145 and 65 million years ago. The name was first applied to extensive deposits of this age that form white cliffs along the English Channel between Great Britain and France.

Crown – the portion of a tooth that usually extends beyond the gums and is usually covered with enamel.

Cursorial – an animal well adapted for running (usually at high speed or over long distances) or for any anatomical feature found in such an animal, such as long, slender limb bones.

D

Deciduous – an anatomical structure, such as a tooth or antler, shed or dropped off the body. Primarily refers to the deciduous tooth series of juvenile mammals, which consists of a set of incisors, canine, and premolars that erupt early in life (sometimes prior to birth) and which are usually shed sequentially as the animal approaches maturity and replaced by members of the adult or permanent tooth series.

Dental arcade – the shape made by the rows of teeth in the upper jaw.

Dental formula – the number of teeth a mammal has of each tooth along one side of the sagittal axis (central line of symmetry). The formula denotes incisors, canines, premolars, and molars in that order. Dental formulas can be written in several ways. The top numbers represent the upper teeth, and the bottom numbers represent the lower teeth (e.g., 2/2-1/1-2/2-3/3 in humans) for 2 sets of incisors, 1 set of canines, 2 sets of premolars, and 3 sets of molars. In some mammals, the number of upper and lower teeth may differ, hence the need to denote top and bottom number of sets.

Dentine (dentin, alt. sp.) – a hard vertebrate tissue similar in composition to bone but found only in teeth and dermal tissue. Dentine exists in several varieties with different internal structures and hardness.

Dentition – a collective term for the teeth of an individual vertebrate or those of a taxonomic group of vertebrates.

Derived – features or suite of features that evolved more recently than other conditions of the feature in question. *Older forms are considered more primitive.*

Derived character state – when comparisons are made between differing character states found on two taxa if one has evolved from (or is thought to have evolved from) the other. The former is considered the derived character state.

Dermal – of or relating to the skin.

Describe – the process by which species are formally defined and given a scientific name. The systematic comparison of specimens of the same and different species and a **description** detailed enough that others can identify the species as well.

Development – the process by which organisms and specific features are formed during gestation and the organism's lifetime *before* reaching maturity.

Devonian – a period of the Paleozoic era between 410 and 360 million years ago; named after Devonshire, England, where rocks of this age were first studied.

Diagenesis – the physical and chemical changes that occur when sediment is converted to sedimentary rock. *All chemical, physical, and biological modifications undergone by a sediment after its initial deposition.*

Diaphysis – the shaft of a mammalian limb bone, the portion that excludes the epiphyses.

Diastema (diastemata, plural) – the gap between consecutive teeth found in some mammals, especially herbivorous groups such as rodents and ungulates.

Digitigrade – the condition of supporting the body weight and walking on the digits or phalanges instead of the metapodials (i.e., *plantigrade*). Dogs and birds are examples of digitigrade animals.

Dinocyst – a resting stage or reproductive stage in the life cycle of a dinoflagellate.

Dinoflagellate – small organisms with both plant-like and animal-like characteristics, usually classified as algae (plants). *Named from their twirling motion and whip-like flagella.*

Diphyodont – any animal replacing its first set of teeth with a second one it retains throughout life.

Dispersal – the action or process of distributing across a wider area or region.

Diurnal – describes organisms whose behaviors occur during the daytime while sleeping at night.

Durophagy (or *durophagous*) – the eating behavior of animals that eat hard-shelled organisms or those with exoskeletons, such as mollusks or crabs.

E

Eocene [Greek, *eos* for dawn and *ceno* for new] – an epoch of the lower Tertiary period between 55.5 and 33.7 million years ago.

Ecology – a branch of biology focusing on the relations between living plants and animals and their environment. *The relationship between organisms to each other and the physical environment.*

Ecosystem – a part of ecology consisting of the environment, its living parts, and the nonliving factors that affect it.

Edentulous – lacking teeth.

Ediacaran (or *Vendian*) – the latest period of the Proterozoic era, between 650 and 544 million years ago. Fossils represent a characteristic collection of complex soft-bodied organisms at several localities worldwide.

Element – part of the skeleton; refers to bones or teeth (e.g., humerus, metacarpal).

Enamel – the hardest vertebrate tissue, composed primarily of the mineral hydroxyapatite. A shiny, glossy substance forms the teeth' outer covering in tetrapods and some fishes.

Enameloid – a variety of very hard dentine that compositionally resembles true enamel and is found on the teeth and scales of some fish. It can only be distinguished from true enamel microscopically.

Endemic –prevalent in or limited to a particular region.

Epiphysis (epiphyses, plural) – separate centers of bone development found at the ends of some bones in juvenile mammals and a few vertebrates.

Epoch – a standard unit of the geologic time scale, epochs are subdivisions of periods. The term is capitalized when it follows a time unit (e.g., *Miocene Epoch*).

Era – a standard unit of the geologic time scale; eras comprise two or more contiguous periods. The term is capitalized when it follows such a time unit (e.g., the Cenozoic Era_.

Estuarine – of an *estuary.*

Estuary – is a body of water where freshwater from rivers or streams flows into the ocean and mixes with seawater. It is considered the transitionary environment between land and sea and is intermediate in salinity between the freshwater of rivers and the seawater of oceans.

Evolution The process of morphologic and genetic change in biological organisms over time that occurs because of natural selection and genetic drift.

Extant – a species or taxon that still exists and has living members, the opposite of *extinct.*

F

Family – the basic rank in the taxonomic hierarchy above the genus and below the order. The formal name of a family must end in the suffix "-idae" which is preceded by the root of the generic name of the genus of the family. For example, Felidae is the formal scientific name of a mammalian family, while felids are its informal equivalent. The species making up a family can be subdivided into subfamilies, tribes, and subtribes. In order of most broadly classified to most specific, the well-known ranks are life, domain, kingdom, phylum, class, order, family, genus, and species.

Fauna – animals of a given region or period of geologic time.

Feral – a once captive or domesticated animal or plant species now living and breeding in the wild, usually in a region where it is not a native species. *For example, the feral burros and horses living in the American Southwest.*

Flora – plants of a given region or period of geologic time.

Fluvial – of a *river*.

Foramen (foramina, plural) – a small hole or opening in a bone, usually for the passage of a nerve or blood vessel.

Foraminifer – protozoans that belong to the subclass Sarcodina, order Foraminifera, which has a test of one-to-many chambers composed of secreted calcite or agglutinated particles.

Formation – the basic unit of rock in the geologic discipline of stratigraphy; it consists of one or more beds of the same type of rock (limestone, sandstone) or alternately a distinctive combination of rock types with a lateral extent of at least several miles. Formal names of formations are capitalized and based on geographic names. *Examples from Florida are the Anastasia Formation and the Suwannee Limestone.*

Fossa (fossae, plural) – a prominent depression or concavity in a bone, often to house a muscle or gland.

Fossils – the recognizable remains of past life on Earth, such as bones, shells, leaves, or other evidence, such as tracks, burrows, or impressions.

Fossilization – the burial of a plant or animal in sediment and the eventual preservation of all, part, or a trace of it.

Fossil record –the number of fossils discovered and the inferred information.

Fossorial – an animal well adapted for digging; or anatomical feature in such an animal.

G

Generalist – organisms with the combination of morphological or behavioral features allowing them to exist in a broader spectrum of environments or exhibit a comprehensive range of behaviors, depending on their environment. Compare *specialists*.

Generalization – the combination of morphological or behavioral features that allows an organism to exist in a broader spectrum of environments or exhibit a wider range of behaviors, depending on its environment. Organisms of this nature are *generalists*. Compare *specialists*.

Generic –of or relating to a genus.

Genus (genera, plural) – the basic taxonomic rank immediately above the species level. Genus consists of a group of species more closely related to each other than any species not included in the genus, but a genus may contain only one species. A formal genus name is capitalized and written in italics, for example, *Canis* or *Triceratops*. Species making up a genus may be grouped into two or more subgenera. In that case, one subgenus will have the same name as the genus, while the others will have different names. Subgeneric names are formally written like generic names but follow the generic name and are enclosed in parentheses.

For example, *Hespertestudo* (*Hesperotestudo*) and *Hesperotestudo* (*Caudochelys*) are two subgenera tortoises found as fossils in Florida. A formal species name has the genus and species name, such as *Smilodon gracilis*, where *Smilodon* is the genus name and *gracilis* the species designation. Note that the second component of a formal species name is meaningless without the genus name. For example, the species designation *gracilis* applies to more than 50 species of animals, plants, and bacteria.

Geochemistry – the science that deals with chemical changes in and composition of Earth's crust.

Geologic time – the period extending from the formation of Earth to the present.

Geologic Time Scale – an arbitrary chronologic sequence of geologic events used to measure the age of any part of geologic time, usually presented as a chart showing the names of the various rock-stratigraphic, time-stratigraphic, or geologic-time units.

Geology – the study of the planet Earth, the materials, the processes acting on these materials, the products formed, and the history of the planet and its life forms since its origin.

Granivore – a herbivorous animal whose diet is entirely or mainly seeds.

Grazer – a herbivorous animal (usually a mammal) whose diet is entirely or mainly grasses.

Group – 1) a basic rock unit in the geologic discipline of stratigraphy consisting of two or more contiguous formations. A group is, therefore, typically a larger and more widespread stratigraphic unit than a formation. *An example from Florida is the Hawthorn Group.* 2) an informal assemblage of species synonymous with taxon.

H

Hadean – the earliest subdivision of the Precambrian spans the formation of Earth, about 4.5 billion years ago, and the start of the Archaean era, 3.8 billion years ago. *This interval predates true geologic time since no rocks of this age are known except a few meteorites.*

Hauterivian – European Lower Cretaceous stage between 122 and 118 million years ago.

Herbivore – an animal whose diet is entirely or predominantly of plant matter.

Heterodont – an animal whose teeth have different shapes and functions. Although most common in mammals, some sharks, fish, and reptiles are heterodonts. Contrast *homodont.*

Holocene [Greek, *holos* for entire and *ceno* for new] – an epoch of the Quaternary period, from the end of the Pleistocene (8,000 years ago) to the present.

Holotype – the single specimen upon which the scientific name of a species is based and was explicitly designated as such in the original published description. All other specimens are only referred to the species.

Homodont – an animal whose teeth all have the same appearance and function. Contrast *heterodont.*

Homonymy (*biological taxonomy*) – two different taxonomic groups have identically spelled scientific names. In such cases, the name published after the other one is invalid and must be replaced.

Hydrogeology – the science of subsurface waters and geologic aspects of surface waters.

Hypercarnivore – an animal whose diet consists of more than 70% meat.

Hypselodont – a tooth with a high-crowned, ever-growing crown lacking roots; or an animal with such teeth. Usually used only in conjunction with mammalian teeth. Found in xenarthrans, rabbits, some rodents, and a few ungulates.

Hypsodont – a tooth with a relatively tall or high crown, especially one in which the crown is taller than the roots, or an animal with such teeth. Usually used only in conjunction with mammalian teeth.

I

Incisor – an anterior-most tooth in the mammalian dentition, usually single-rooted, with either a chisel-shaped or conical crown.

Incisiform – a tooth resembling a typical mammalian incisor, especially regarding the lower canine in some artiodactyls.

Index fossil – in biochronology, describes a species or genus whose chronologic range is restricted to a particular biochronologic unit. For example, the horse species *Neohipparion eustyle* is an index fossil for the Hemphillian land mammal age. It occurs only in Hemphillian fossils sites and does not occur at sites of any other age.

Intermediate feeder (or mixed-feeder) – a herbivorous animal exhibiting adaptations for grazing and browsing.

Interspecific – relating to characteristics found in *two or more* species.

Intraspecific – relating to characteristics found within a *single* species.

Isotopes – atoms with the same number of protons but different numbers of neutrons (i.e., isotopes).

Isotopic dating (or *radiometric dating*) – age determination methods based on nuclear decay of naturally occurring radioactive isotopes. Age in years for geologic materials is calculated by measuring the presence of a short-life radioactive element (e.g., carbon-14 or ^{14}C) or by measuring the presence of a long-life radioactive element plus its decay product (e.g., potassium-40/argon-40)

J

Jurassic – the middle period of the Mesozoic era, between 213 and 145 million years ago; named after the Jura Mountains between France and Switzerland, where rocks of this age were first studied.

K

Keratin (or **keratinous**) – hard, organic dermal tissue that makes up fingernails and the sheaths of horns; on occasion, it preserves in fossils.

L

Labial (or *buccal*) – the external (or lateral) side of a tooth, towards the lips or cheek.

Land mammal ages – refers to the North American land mammal ages (NALMA), a timescale established for prehistoric fauna in North America using geographic place names for where the initial fossils were collected.

Life history – the series of changes an organism undergoes during its lifetime, the phases of which are typically marked by biological phenomena, such as reaching sexual maturity.

Lingual – refers to the internal side of a tooth, towards the tongue. Contrast *labial side*.

Lithologic unit (or *lithostratigraphic unit*) – a body of rock consistently dominated by a certain lithology or similar color, mineralogic composition, and grain size. It may be igneous, sedimentary, or metamorphic and may or may not be consolidated.

Locomotion – the ability to move from one place to another. Different forms of locomotion produce different morphologies, which is of interest to morphologists.

Lophodont (or *lophs*) – describes a tooth in which ridges connect the cusps or cones on the crown. Usually refers to mammalian teeth.

M

Maastrichtian – European stage of Upper Cretaceous, between 72 and 66 million years ago.

Macrofossil – a fossil large enough to be studied without a microscope.

Mandibular symphysis – a vertical ridge that marks the fusion of the right and left parts of the mandible.

Marine – of the ocean or saltwater environments.

Matrix – samples of fossil-bearing sediment to be processed by screen-washing.

Member – a basic unit of rock in the geologic discipline of stratigraphy consisting of a portion of a formation. The subdivision of a form into two or more members is optional. For example, the Arcadia Formation includes beds placed in the Tampa Member and the Nocatee Member in southern Florida.

Mesic – a region or habitat with ample rainfall and vegetation.

Mesozoic [Greek for *middle life*] – an era of geologic time between the Paleozoic and the Cenozoic, 248 and 65 million years ago.

Metacarpals – the metapodials of the *hand* or *forelimb.*

Metapodial – the set of bones in the limbs of tetrapods found between the carpals or tarsals and the phalanges.

Metatarsals – the metapodials and those of the *hind limb.*

Microfossil – a small fossil that must be studied with a microscope.

Micrometer – a unit of measure; one million micrometers equals one meter.

Microwear –microscopic patterns of damage on the surface of teeth or bone.

Mineral – the inorganic, crystalline compounds that make up rocks. Examples are *quartz*, *calcite*, *gypsum*, and *feldspar*. The mineral found in the vertebrate skeletal system is hydroxyapatite, a phosphatic mineral synthesized by special bone-producing cells.

Miocene [Greek, *meion* for less and *ceno* for new] – an epoch of the upper Tertiary period, between 23.8 and 5.3 million years ago.

Mississippian – a period of the Paleozoic era between 360 and 325 million years ago; named after the Mississippi River valley, which contains good exposures of rocks of this age.

Mixed-feeder – a herbivorous animal (usually a mammal) whose diet is grasses and leaves.

Molar – one of the posterior-most series of mammalian teeth, usually with more than one root and a complicated occlusal surface with numerous cusps and cones. By definition, a molar does *not* have a *deciduous precursor*.

Molariform – a tooth appearing as a typical mammalian molar, especially for the posterior premolars in many herbivorous groups.

Monogamy (or *pair bonding*) – not to be confused with the socio-cultural definition of monogamy, typically refers to the behavior exhibited by animals that pair-bond for some time, sometimes for life. It is a strategy used by animals who exhibit biparental care. However, copulations outside of the pair-bonded unit are observed frequently in nature in numerous species of animals, ranging from birds to primates.

Monophyletic – a taxonomic group that includes the closest common ancestor of the group and all its descendants.

Morphology – the form of a skeleton or its parts and its structure. *Studying the form and structure of animals and plants and their fossil remains.*

N

Nektonic – describes aquatic organisms that swim.

Neoteny – the retention of juvenile features in an adult animal.

Neotype – a specimen formally designated in a published study to replace a missing or destroyed holotype. After a neotype has been designated, it has the same importance and duties as a holotype.

Node – the point at which branches of a phylogenetic tree diverge; representative of a common ancestor between diverged branches.

Notochord – a rodlike cord of cells in lower chordates that forms the main lengthwise support structure of the body.

O

Occlusal – relating to or involving the surface of teeth where they touch each other, as in chewing or biting down.

Occlusion – the relationship between the upper and lower teeth when they come into contact.

Oligocene [Greek, *oligo* for little or few and *ceno* for new] – an epoch of the early Tertiary period, between 33.7 and 23.8 million years ago.

Omnivores – animals that eat both plant and animal material.

Ontogenesis (or *ontogenetic*) – the development of an individual or anatomical or behavioral feature from its earliest stages to maturity.

Ontogeny – the study of *ontogenesis*.

Orbit – the depression or deep concavity in the skull which houses the eye.

Order – taxonomic rank used in the classification of organisms. In order of most broadly classified to most specific, the well-known ranks are life, domain, kingdom, phylum, class, order, and family. Members within a particular order are evolutionarily more closely related to each other than any other order within their class.

Ordovician – the second earliest period of the Paleozoic era, between 505 and 440 million years ago; named after a Celtic tribe called the Ordovices.

Osteoderm (or informally a *scute*)– a piece of bone found in the skin of some vertebrates. Especially well-developed in crocodilians, tortoises, and armadillos.

Osteological – of or relating to bones.

Ostracoderms – primitive jawless fishes covered by bony armor from Cambrian through Devonian periods.

Otolith – small, oval structures made of calcium carbonate that grow in the inner ear region of some fishes. Fossil otoliths are typically only studied by specialists.

P

Pair bonding (or *monogamy*) – not to be confused with the socio-cultural definition of monogamy, typically refers to the behavior exhibited by animals that pair-bond for some time, sometimes for life. It is a strategy used by animals who exhibit biparental care. However, copulations outside of the pair-bonded unit are observed frequently in nature in numerous species of animals, ranging from birds to primates.

Paleobathymetry – the study of ocean depths and topography of the ocean floor in the geologic past.

Paleobiogeography – the branch of paleontology that deals with the geographic distribution of plants and animals in past geologic times, especially regarding ecology, climate, and evolution.

Paleoceanography – the study of oceans in the geologic past, including their physical, chemical, biological, and geologic aspects.

Paleocene [Greek, *palaois* for old and *ceno* for new] – the earliest epoch of Tertiary period, between 65 and 55.5 million years ago.

Paleoclimate – the climate of a given period in the geologic past.

Paleoecology – the inferred relationship between past organisms to each other and their past physical environment. *The study of the relationships between ancient plants and animals and their environments.*

Paleoenvironment – environment of the geologic past.

Paleontologists – scientists who study fossils.

Paleontology – the study of life in past geologic times. *Fossilized animals and plants.*

Paleozoic [Greek for *old life*] – an era of geologic time from the end of Precambrian to the beginning of the Mesozoic, between 544 and 248 million years ago.

Parallel evolution (*convergence* or *parallelism*) – the independent acquisition of the same derived character state in two or more taxonomic groups not through a common ancestor.

Parallelism (*convergence* or *parallel evolution*) – the independent acquisition of the same derived character state in two or more taxonomic groups not through a common ancestor.

Paraphyletic – a taxonomic group that includes the closest common ancestor but not all of the descendants of that species. Paraphyletic groups are not considered valid by some scientists.

Paratype – one or more referred specimens designated explicitly in the original description of a new species that provide additional information beyond the *holotype*.

Parsimony (or *parsimonious*) – in systematics, the acceptance of one possible phylogenetic arrangement in deference to other arrangements because it requires the fewest number of character state reversals and parallelisms.

Pathology (or *pathological*) – an individual or anatomical part of an individual with an unusual morphology caused by illness, injury, malnourishment, or genetic defect.

Pectoral – relating to the anterior paired lateral fin in chondrichthyan and osteichthyan fish, the forelimb in tetrapods, or the region where the forelimb attaches to the torso.

Pectoral girdle – the bones in the shoulder region in tetrapods that articulate with the proximal end of the humerus and anchor the forelimb to the torso structurally and as attachment sites for many of the muscles that move the forelimb. Individual pectoral girdle elements include the scapula, coracoid, and clavicle.

Pelage – the hair, fur, or wool of a mammal.

Pelagic – refers to open water marine habitats free of direct influence of the shore or ocean bottom. Pelagic organisms are generally free-swimming (*nektonic*) or floating (*planktonic*).

Pelvic girdle – bones in tetrapods that articulate with the proximal end of the femur and anchor the hind limb to the torso structurally and as attachment sites for many muscles moving the hind limb. Individual elements of the pelvic girdle are the *ilium*, *ischium*, and *pubis;* in adult mammals, these three elements fused to form a single bone called the *pelvis* or *innominate*.

Pennsylvanian – a period of the Paleozoic era, between 325 and 286 million years ago; named after Pennsylvania, where rocks of this age are widespread.

Period – a formal unit of the geologic time scale, a subdivision of time between the ranks of era and epoch. The term is capitalized when it follows the formal name of such a time unit, for example, the Cretaceous Period.

Periotic (or *petrosal*) – the mammalian skull bone containing the inner ear. Crucial evolutionary information includes these and other features of the skull's ear region.

Permian – the final period of the Paleozoic era, between 286 and 248 million years ago; named after the province of Perm, Russia, where rocks of this age were first studied.

Petrosal (or *periotic*) – the mammalian skull bone that houses the inner ear. Detailed analysis of these and other features of the ear region of the skull often reveals crucial evolutionary information.

Phalanx (**phalanges**, plural) – one of the bones in the digits (fingers and toes) of a tetrapod.

Phanerozoic – the period (or *eon*) between the end of the Precambrian and today. *Phanerozoic begins with the start of the Cambrian period, 544 million years ago. It encompasses the period of abundant, complex life on Earth.*

Phenetic (or *taximetrics*) – the classification of organisms based on overall similarity in morphology, regardless of phylogeny or evolutionary relation.

Phylogeny – the pattern of ancestor-descendant relationships among a group of biological species deduced by studying character states.

Piscivory (or *piscivorous*) – an animal whose diet primarily consists of fish.

Placoderms – a group of primitive armored jawed fish found almost exclusively in rocks from the Devonian Period.

Plankton – aquatic organisms that drift or swim weakly.

Planktonic – describes aquatic organisms that float.

Plantigrade – the condition of supporting the body weight and walking on the metapodials instead of the phalanges (i.e., *digitigrade*). Humans and bears are examples.

Plastron – the ventral portion of the shell of a turtle.

Pleistocene [Greek, *pleitos* for most and *ceno* for new] – an epoch of the Quaternary period, between 1.8 million years ago and the beginning of the Holocene 8,000 years ago.

Pliocene [Greek, *pleion* for more and *ceno* for new] – the final epoch of the Tertiary period, between 5.3 and 1.8 million years ago.

Polygyny [*poly-* meaning 'many' and *-gyn* meaning 'female'] – a social system where one or few males control reproductive access to most females within a social unit. Sexual dimorphism is characteristic of species that exhibit **polygynous** behavior because of increased competition between males for mating access to females.

Polyphyletic – a taxonomic group that does not include the closest common ancestor of the group. Polyphyletic groups are not considered valid by scientists.

Polyphyodont – any animal whose teeth are continuously replaced.

Postcranial – of or related to that portion of the skeletal system other than the skull and jaws.

Precambrian [*before Cambrian*] – all geologic time before the beginning of the Paleozoic era; includes about 90% of geologic time from the beginning of Earth, about 4.5 billion years ago, to 544 million years ago.

Prehensile – capable of grasping, such as a prehensile tail in some species of New World monkeys.

Premolar – one of the mammalian teeth posterior to the canine and anterior to the molars, usually with more than one root. They vary in morphology from simple to complicated and molariform depending on the type of mammal. Premolars usually have a deciduous precursor, but in some cases, the deciduous premolar is retained in the adult.

Preorbital – anatomically located anterior to the orbit.

Preparation (*prep work* or *prepping*) – in paleontology, the surrounding rock or sediment is systematically removed from fossils, broken parts are glued back together, and weak areas are strengthened with consolidants. Most fossils require some degree of preparation for future analysis and use in studies and long-term storage or display. A person trained in these activities is called a *preparator*.

Primitive – a feature or suite of features that exhibit an older condition than seen in more recent representatives of a particular lineage. Those exhibiting recent conditions of a feature are considered *derived*.

Primitive character state – when comparisons are made between two different character states, if one has evolved from (or is thought to have evolved from) the other, the latter is considered the primitive character state.

Prospect – search for potentially fossiliferous deposits in a particular area.

Proterozoic [early life] – the final era of the Precambrian, between 2.5 billion and 544 million years ago. Fossils of primitive single-celled and more advanced multicellular organisms appear in abundance in rocks from this era.

Protist – an organism that belongs to the kingdom Protista, which includes forms with both plant and animal affinities (i.e., protozoans, bacteria, some algae, fungi, and viruses).

Q

Quaternary [Latin, *four at a time*] – the second period of the Cenozoic era spanning 1.8 million years and now; contains two epochs: the Pleistocene and Holocene.

R

Radiometric dating (or *isotopic dating*) – age determination methods based on nuclear decay of naturally occurring radioactive isotopes. Age in years for geologic materials is calculated by measuring the presence of a short-life radioactive element (e.g., carbon-14) or by measuring the presence of a long-life radioactive element plus its decay product (e.g., potassium-40/argon-40)

Ramus (**rami**, plural) – one side of a mammalian mandible. Technically, "mandible" refers to the combined right and left rami but is often used informally to indicate only one side.

Reworked – fossils that eroded from the sediments or rock enclosing it were deposited in younger sediments. In this manner, fossils of different ages can be found in the same rock unit. Reworked fossils can often be recognized by their worn appearance.

Rhizopod – a protozoan of the class Rhizopoda that has pseudopodia.

Root – the basal portion of a tooth (as opposed to enamel-covered crown) where it attaches to the jaw.

Rostral – relating to the *rostrum*.

Rostrum (**rostra**, plural) – the anterior portion of the skull, usually consisting of the maxillary, premaxillary, and nasal bones.

Rudist – an extinct bivalve mollusk from the Jurassic and Cretaceous with two different sized and shaped shells; they usually were attached to the substrate and were either solitary or in reeflike masses.

S

Sacral – of or relating to the vertebrae that articulate with the pelvis. In some vertebrates, these vertebrae are fused to form a *sacrum*.

Sacrum – fused vertebrae that articulate with the pelvis.

Salinity – the saltiness or dissolved salt content of a body of water.

Sample – a part or quantity used to represent the whole.

Santonian – European stage of the Upper Cretaceous between 88 and 84 million years ago.

Scanning Electron Microscope (SEM) – a microscope in which a finely focused beam of electrons is scanned across a specimen. *Electron intensity variations are used to construct an image of the specimen and ideal for magnifications from 200 to 35,000.*

Scansorial – 1. animals well adapted for climbing, 2. anatomical features in such an animal.

Scavenger – an animal that feeds on decaying carcasses, dead plant material, or garbage.

Screen-wash – the process of separating fossils from sediment using running water and one or more sets of boxes with bottoms of different-sized screens. Sedimentary particles smaller than the screen openings wash through, leaving a residue of concentrated grains and fossils sorted after drying. *Screen-washing collects fossils of small rodents, birds, lizards, snakes, salamanders, frogs, and fish.*

Scute (informal or *osteoderm*) – a piece of bone found in the skin of some vertebrates. Especially well-developed in crocodilians, tortoises, and armadillos.

Sediment – solid unconsolidated rock and mineral fragments that come from the weathering of rocks and are transported by water, air, or ice and form layers on Earth's surface. Sediments can also result from chemical precipitation or secretion by organisms.

Sedimentary rock – a rock resulting from the consolidation of sediments.

Selenodont – a type of lophodont tooth in which crescentic ridges connect the cusps or cones. Characteristics of ruminant and tylopod artiodactyls such as bison and deer.

Serrated – having a notched or jagged edge like the cutting edge of a saw. In paleontology, often used to describe teeth. *For example, the canine teeth of some saber-toothed cats had serrated edges, as do the teeth of some species of sharks.*

Sesamoid – a bone that forms in a joint capsule or tendon near a joint. In many medium- to large-sized mammals, there are well-formed sesamoids found in joints between metapodials and the proximal-most phalanges. The patella (or kneecap) is typically the largest sesamoid in the body.

Sexual dimorphism – when the two sexes of a species differ in size or morphology of one or more characters.

Silurian – a period of the Paleozoic spanning between 440 and 410 million years ago; named after a Celtic tribe called the Silures.

Sister taxon (**sister taxa**, plural) – a species or group that shares its most common ancestor with one or more other species or groups (e.g., sister clade, sister family). Sister taxa are more closely related to each other than any other taxon.

Specialization – the presence of morphological or behavioral features with a particular and narrow-ranged function. The term *specialized* is often used in a relative sense to describe groups or species where one is more specialized than the other, but both can be specialized when compared to an outgroup that is more *generalized*. Organisms that are specialized are *specialists* and exhibit *specializations*.

Species – the lowest commonly used rank in the taxonomic hierarchy. A species is a group of morphologically similar individuals with more diagnostic character states distinguishing them from all other species. Some taxonomists define a species as a group of interbreeding individuals, although this is difficult to test in modern animals and impossible in fossils. A formal species name consists of two words written in italics, for example, *Homo sapiens* or *Tyrannosaurus rex*. Note that without the genus name, the second component of a formal species name is meaningless. For example, the species designation *gracilis*, as in *Smilodon gracilis*, applies to more than 50 species of animals, plants, and bacteria.

Stable isotopes – atoms with the same number of protons but different numbers of neutrons (i.e., isotopes) that do not decay into other elements over time. *Unlike stable isotopes, radioactive isotopes are unstable and decay into other elements over time.*

Stable isotope analysis – uses stable isotopes to learn about the chemical (elemental) makeup of the past, such as determining diets, environments (different plants produce isotopic signatures), and climate.

Stratigraphy – the branch of geology concerned with the formation, composition, ordering in time, and arrangement in space of sedimentary rocks. *The study of the order and relative position of strata and their relationship to the geologic time scale.*

Stratum (**strata**, plural) – a layer or series of rock in ground denoting a period when sediment was being deposited.

Subfamily – taxonomic classification rank denotes a level below *family* but falls above *genus* level. Used when genera within a family can be divided into subfamilies based on morphological or molecular similarities that groups them phylogenetically from one another. In order of most broadly classified to most specific, the well-known ranks are life, domain, kingdom, phylum, class, order, family, genus, and species.

Subfossil – a fossil less than 10,000 years old.

Subspecies – the lowest formal taxonomic rank, subdivisions of a species, used to separate different populations within a species, usually on criteria such as color patterns or size. A formal subspecies name consists of three words written in italics, for example, *Lynx rufus koakudsi*. Subspecies are infrequently used in vertebrate paleontology but commonly used among ecologists who want to differentiate a particular population within a species from another population within that species.

Subtropical – bordering on the tropics or nearly tropical.

Supraorbital – anatomically located dorsal to the orbit.

Symphysis (**symphyses**, plural) – immovable (or nearly so) joint where a skeletal element's right and left sides fuse. Examples are the mandibular symphysis and the pelvic symphysis.

Synonymy – in biological taxonomy when a single taxonomic group has two or more scientific names. The original name is a *senior synonym*, while all other names given after the first name are *junior synonyms*. When species names are *synonymized*, all specimens under multiple names are referred to by the original scientific name given to the first described fossil.

Systematics – the scientific study of the phylogeny of biological organisms and the methods used to reconstruct their evolutionary relationships.

T

Talus (or *astragalus*) – bone articulating with the tibia and fibula to form the ankle joint.

Tarsal – any skeletal elements of the ankle, such as the cuboid or astragalus. Distally, the tarsal bones articulate with the metatarsals, proximally with the tibia and fibula.

Tarsometatarsus – the fusion of the metatarsals, and some tarsals, forms a single bone.

Tarsus – collectively, the skeletal elements of the ankle, such as the cuboid or astragalus. Distally, the tarsal bones articulate with the metatarsals, proximally with the tibia and fibula.

Taximetrics (or *phenetics*) – the classification of organisms based on overall similarity in morphology, regardless of phylogeny or evolutionary relation.

Taxon (taxa, plural) – any organism group comprising one of the scientific classification hierarchies. For example, the genus *Canis*, the family Bovidae, and the class Actinopterygii are all taxa.

Taxonomy – the scientific study of classifying and naming biological organisms. *The science of identifying, naming, and classifying plants and animals.*

Terrestrial – refers to land habitats in distinction from water (aquatic) habitats.

Tertiary – the first period of the Cenozoic era (after the Mesozoic era and before the Quaternary period), between 65 and 1.8 million years ago.

Tetrapod – a common name for the taxon Tetrapoda, which includes amphibians, reptiles, birds, and mammals. *Most tetrapods live on land, although some are secondarily aquatic.*

Thoracic – relating to the chest region (the thorax), especially the vertebrae in a mammal that articulates with a rib.

Trace fossil – a fossil that consists of something made by an ancient organism while it was alive. *Fossilized footprints, burrows, nests, stomach stones, and coprolites are examples of trace fossils.*

Triassic – the earliest period of the Mesozoic era, between 248 and 213 million years ago. Triassic refers to the threefold division of rocks of this age in Germany.

Tribe – s taxonomic rank above genus but below family and subfamily and is sometimes subdivided into subtribes.

Trilophodont – a tooth primarily composed of three transverse ridges. Usually used only in conjunction with mammalian teeth, especially proboscideans.

Trophic level – any of several hierarchical levels in a food chain where members of each trophic level serve the same function in that food chain.

Tropical – refers to climatic conditions like those in the region on Earth between the Tropic of Cancer and the Tropic of Capricorn; it includes high temperature and humidity and abundant rainfall.

Turonian – European stage of the Upper Cretaceous, between 91 and 90 million years ago.

Tympanic bulla (or *auditory bulla*) – the bone that encloses the inner ear region of the mammalian skull and is found on the ventral surface of the posterior region of the skull. *Typically, fragile, and thus rarely preserved in fossils, but inflated and composed of solid bone in whales. For that reason, whale auditory bullae are commonly recovered as fossils.*

Type genus – defines a biological family and is the root of the family name in taxonomy.

Type locality – fossil site or locality where the holotype specimen of a species was collected.

Type species – the genus or subgenus, under modern rules of scientific names, must explicitly designate when a new genus or subgenus is formally named. If a scientist determines that the species placed in a single genus belong to two or more genera, then the group of species that includes the type species will remain in the original genus; the other species are assigned to a different genus.

Type specimen – a formally selected specimen intended to serve as the principal reference for a particular species. Used to help identify that species in the fossil record. See *holotype*.

U

Ungual – relating to a hoof, claw, or nail.

Ungulate – a hoofed mammal.

Unguligrade – an extreme form of digitigrade posture where the body weight is supported by only the distal-most phalanx of one or a few digits. Fairly common in artiodactyls such as bison and deer, but otherwise rare among other mammals.

V

Valanginian – European Lower Cretaceous stage spanning 131 and 122 million years ago.

Variation – the differences in size, color, morphology, behavior, and other biological characteristics between two or more individuals.

Vendian (or *Ediacaran period*) – the latest period of the Proterozoic era, between 650 and 544 million years ago. It is distinguished by fossils representing a characteristic collection of complex soft-bodied organisms at several localities worldwide.

Vertebrate – a common name for a member of the monophyletic taxon Vertebrata, which includes jawless fishes, bony fishes, sharks and rays, and tetrapods. Almost all vertebrates have a series of bones known as vertebrae (hence the name) and can produce calcium phosphate crystals. Among animals, vertebrates uniquely possess a brain that coordinates the nerves and senses and is protected and supported by a chondrocranium.

Volant – an organism capable of powered flight, like most birds and bats.

X

Xeric – a dry habitat characterized by low levels of rainfall, such as an arid or semiarid region.

Frank J. Addivinola, Ph.D.

The lead author and chief editor of this study guide is Dr. Frank Addivinola. With his outstanding education, laboratory research, and decades of university science teaching, Dr. Addivinola lent his expertise to develop this book.

Dr. Frank Addivinola conducted original research in developmental biology as a doctoral candidate and pre-IRTA fellow in Molecular and Cell Biology at the National Institutes of Health (NIH). His dissertation advisor was Nobel laureate Marshall W. Nirenberg, Chief of the Biochemical Genetics Laboratory at the National Heart, Lung, and Blood Institute (NHLBI). Before NIH, Dr. Addivinola researched prostate cancer in the Cell Growth and Regulation Laboratory of Dr. Arthur Pardee at the Dana Farber Cancer Institute of Harvard Medical School.

Dr. Addivinola holds an undergraduate degree in biology from Williams College. He completed his Masters at Harvard University, Masters in Biotechnology at Johns Hopkins University, and five other graduate degrees at the University of Maryland University College, Suffolk University, and Northeastern University.

During his extensive career, Dr. Addivinola held faculty positions at colleges and universities, including Harvard University, Johns Hopkins University, University of Maryland and Northeastern University, and taught numerous undergraduate and graduate-level courses in biology, biochemistry, organic chemistry, inorganic chemistry, anatomy and physiology, medical terminology, nutrition, and medical ethics. He received several awards for his research and presentations.

Essential Physics Self-Teaching Guides

Kinematics and Dynamics

Equilibrium and Momentum

Force, Motion, Gravitation

Work and Energy

Fluids and Solids

Waves and Periodic Motion

Light and Optics

Sound

Electrostatics and Electromagnetism

Electric Circuits

Heat and Thermodynamics

Atomic and Nuclear Structure

Visit our Amazon store